# GOD'S STEWARDS

# God's Stewards

## *The Role of Christians in Creation Care*

**Edited by Don Brandt**

**Foreword by Eugene H. Peterson**

**World Vision**

A Policy & Advocacy Report.
Produced by World Vision Publications on behalf of the World Vision Partnership.

Printed in the United States of America.

09 08 07 06 05 04 03 02     5 4 3 2 1

Published by World Vision International, 800 West Chestnut Avenue, Monrovia, California 91016-3198, U.S.A.

Library of Congress Cataloging-in-Publication Data
God's stewards : the role of Christians in creation care / edited by Don
Brandt ; foreword by Eugene H. Peterson.
     p. cm.
Includes bibliographical references.
 ISBN 1-887983-42-2 (alk. paper)
 1. Ecology—Religious aspects—Christianity.  I. Brandt, Don.
 BT695.5 .G63 2002
 261.8'362—dc21
                         2002008518

Senior editor: Rebecca Russell. Copyediting: Pamela Martin and Bob Newman. Typesetting and cover design: Richard Sears. Cover photo: Todd Bartel/World Vision.

✿ This book is printed on acid-free recycled paper.

# Contents

# Contents

# Contributors

**The Reverend Peter Harris** is an ordained Anglican (Episcopalian) minister with an abiding interest in the care of creation that points people to the redemptive work of Christ. He currently serves as International Director of A Rocha, a Christian environmental organization that he helped found. Together with his wife Miranda, Harris worked for twelve years in Portugal to establish and direct A Rocha's first field study center. His joys and struggles in this adventure are documented in his book, *Under the Bright Wings*. There are now eight A Rocha projects around the world. For the last five years, Harris has been living in France with his family as part of the national team while coordinating A Rocha projects elsewhere. (See www.arocha.org.)

**Dr. R. J. (Sam) Berry** was Professor of Genetics at University College London from 1978 to 2000. His distinguished career as a scientist includes time as President of the Linnean Society and membership in the British Ecological Society and the European Ecological Federation. Dr. Berry has long been involved in issues concerning faith and science. In recognition of his leadership in these subjects, Dr. Berry currently chairs the Environmental Issues Network of Churches Together in Britain and Ireland. In 1996, he was awarded the Templeton U.K. Prize for "long and sustained advocacy of the Christian faith in the world of science." Dr. Berry's most recent books include *God and the Scientist; Science, Life and Christian Belief* (with Malcolm Jeeves) and an edited volume, *The Care of Creation*. His Gifford Lectures given at the University of Glasgow will soon be published as *God's Book of Works*.

**Dr. Michael S. Northcott** is Reader in Christian Ethics at the University of Edinburgh and Academic Chair of the Centre for Human Ecology, an ecological think tank in Scotland. He also pastors a congregation in the dockland area of Edinburgh. Dr. Northcott served as visiting professor at Dartmouth College and the Claremont School of Theology. He has traveled widely in southeast Asia where he visited tribal groups involved in the struggle against commercial logging of the rainforest. His experiences there figure strongly in his passion for connecting Christian faith with environmental issues. Dr. Northcott is the author of more than thirty scholarly papers and of four books including *The Environment and Christian Ethics, The Urban Theology Reader* and *Life After Debt: Christianity and Global Justice.*

**Dr. Anne M. Clifford, C.S.J.** is Associate Professor of Theology at Duquesne University and a member of the Sisters of St Joseph. The major focus of her research is the relationship of Christian theology to the natural sciences. Dr. Clifford is the author of *Introducing Feminist Theology* and more than a dozen articles including "Foundations for a Catholic Ecological Theology of God," in *And God Saw That It Was Good: Catholic Theology and the Environment.* She served as a consultant to the U.S. Catholic Conference's Environmental Justice Program and the National Conference of Catholic Bishops' Committee on Science and Human Values. Among Dr. Clifford's honors are a Duquesne University presidential award for "Excellence in Teaching" and a John Templeton Foundation Science and Religion Course Award.

**Dr. Donald P. Brandt** has been on the staff of World Vision (WV) for 17 years. He is engaged in a variety of socioeconomic-political issues that directly impact WV and may affect the church. Prior to coming to WV, Dr. Brandt spent 15 years as a university professor, during which time he helped direct an interdisciplinary program in environmental studies. Most of Dr. Brandt's research has been published as WV internal documents. External articles have appeared in the *Journal of Humanitarian Assistance* and *Missiology.*

# Foreword

I grew up in one of the more beautiful places of the earth: a valley in the Rocky Mountains of Montana. It was a small town surrounded by foothills and then, ten or fifteen miles away, bounded by the great glacially carved Whitefish and Swan and Mission ranges of the Rocky Mountains. As a boy of nine or so, I began a practice that I have continued with variations ever since. I rode my bike to the base of the foothills and, with my lunch in a daypack, spent wonderful hours immersed in the forest and hills and weather. The Kootenai and Salish Indians had roamed this valley and these hills not long before I was born and I took a great deal of care not to be seen by them. The country was home to bears and mountain lions; but my imagination, well-stocked with stories of David shepherding his flock in the wilderness, gave me the confidence that I could take care of myself in their company quite as well as he did.

But I grew up in a church that took no interest in such things. We were preoccupied with things spiritual. "Spiritual" was understood exclusively in contrast to the material and the physical. We were headed for heaven and anything previous to or other than heaven was a distraction. In later years, I heard someone describe people like us as preoccupied with "the geography of heaven and the temperature of hell." That's certainly what it seemed like to me. The streets paved with gold in the new Jerusalem got far more attention than the forest paths along the old Stillwater River strewn with copper leaves in the autumn.

The heaven and earth that God had joined together had been, in the pungent biblical phrase, "put asunder" in the worshiping community in which I grew up—there was simply no connection between them. But I was in the somewhat confused state of loving them both. I loved the mountains and streams, the fragrance of fresh-leafing cottonwoods in the spring. And I loved the wonderfully storied preaching and robust singing in our little congregation. They were different worlds inhabited by different people speaking different languages. I felt equally at home in both of them, but I experienced little, if any, connection between them.

As I grew up, I continued to maintain my citizenship in these two worlds—the world of creation and the world of Christ. There have been occasional and most blessed exceptions, but for the most part the people in the two worlds have maintained an aloofness from each other. My childhood church may have been a little bit extreme but, compared to the churches I have been part of in my adulthood, not so very much. For the most part still, the people I meet and read who love the creation keep their distance from the church; and the people who love Jesus stay clear of those they dismissively label as "New-Age tree-huggers." The nature people show up in church occasionally for baptisms and marriages and funerals. There are times, after all, when the church can provide a useful service; it's there to be used when needed, but there is no appreciation or love of the church of Christ for its own sake. And the church people show up on a lake or mountain for occasional recreation or "to get away from it all." And sometimes they use canoeing or hiking or water-skiing in the outdoors as a tool for evangelism. But there is no love of the place for its own sake.

From time to time, I wonder why there is so little cross-fertilization between the people who love mountains and birds and trees and the people who pray to the Father and Son and Holy Spirit. How did we get separated into these two camps—the nature people and the Christ people?

It's beyond my competence to trace and analyze how this disconnection took place, but I hope it's not beyond my competence to do something about it. What I have done, in as many ways as I have been able to find and using the Scriptures as my text, is to insist that there is no disconnection in our biblical revelation, no division between creation and Christ, no separation in the evangelical gospel between heaven and earth. Along the way I have been pleased to find allies who are giving intelligent and prayerful witness in their work and writings to the essential unity of "all things" created by the "first-born of all creation" (Col.1:15-16, RSV). The essays in this collection continue to extend the circle of friends in Christ and in creation.

Eugene H Peterson
Professor Emeritus of Spiritual Theology
Regent College (Canada)

# Introduction

*G*od's Stewards: The Role of Christians in Creation Care is written primarily as a call to Christians, by Christians, to live out the Bible's insistence on faithful creation care. This book, though, focuses on a topic that should be of utmost importance to all people, including followers of other faiths and those of more secular inclinations: the biosphere—our environment and home. The biosphere (bios) is a comprehensive term for all the created. It is God's creation and includes all flora and fauna, all organic and inorganic objects, including humankind. Christianity teaches that while people are created "in the image of God," we are nonetheless a species created in a dependent relationship with the Creator.

Reflecting on our role in the created order is not a common practice among Christians. In fact, the attitude of many Christians towards the environment may be characterized as indifference or complacency. In this respect, Christians are probably no different than people of other beliefs. Shocked by statements that raise Mother Earth to a goddess, some Christians shun any environmental or ecological movement. A central belief of other Christians is that they are just "passing through," and that if the end times are drawing near, saving the environment is irrelevant and depletes energy and resources from saving souls.

Many environmental heresies among Christians, as well as criticisms hurled at Christians by secular environmentalists, deep ecologists, eco-feminists and others arise from a misunderstanding of the Genesis story of creation. When placed in the context of the whole Bible, English words like "multiply," "dominion," "subjugate" and "naming" do not depict humankind as an environmental autocrat. Instead, a picture emerges of a created species with responsibilities to care for and bless the rest of creation.

If humans are God's caretakers of creation, then Christians should be on the forefront of movements to protect and renew the environment.[1] While we believe this is a God-given duty, humankind will probably reap the ultimate benefits of not "fouling our nest" more than will other creatures. It's difficult to conceive how sustainable development, "the ability to meet the needs of the present without compromising the ability of future generations to meet their own needs,"[2] can be accomplished without an attitude of being good stewards of God's creation.

Environmental issues are important to World Vision, a Christian development, relief and advocacy partnership that serves more than 85 million people in more than 80 countries. World Vision's mission statement embodies a commitment to follow Christ's example by working with the poor and oppressed in pursuit of justice and human transformation. In carrying out its ministry, World Vision has tended to give ecological matters a passive, albeit a supportive, role in its work. Projects designed to increase food productivity have invariably led to work to abate soil erosion and water run-off, increase soil nutrients, and grow food supplements. Often these programs depend on the introduction of better organic farming and agro-forestry systems and techniques. Environmental renewal isn't the motivating reason for rural projects, but World Vision staff recognize that renewing an abused bios can significantly reduce incidences of abject poverty.

As an organization, World Vision is committed to sustainable development.[3] With this there is the recognition that for development to be "sustainable," the environment must be, too. In this respect, staff working in community projects have been pioneers. They have been leagues ahead of other World Vision personnel, who are often not intimately aware of what's happening "on the ground." In this regard, this book is addressed to the World Vision Partnership, to the global Christian church and to the interested reader.

Reflections by Christians towards the environment, such as those expressed in this book, were quickened by preparations for the World Summit on Sustainable Development Conference (Rio + 10). Chances are that environmental concerns will remain high on the agenda of many churches, civil groups and governments during the months that precede and follow the conference. Statements made will echo positions by which agencies and organizations want to be identified. Other groups will issue reports that indicate the directions in which the organization would like to move. An example is this book, commissioned and edited by the Policy and Advocacy Department of World Vision International.

While each chapter in this book is an individual expression of an author's beliefs and convictions, there are a number of recurrent themes.

- ❧ Christianity is neither anti-environmental nor anthropocentric (human-centred) in its view of creation and the bios.

- ❧ Critics of the biblical Genesis stories have a mistaken notion of the Hebraic meaning of such words as "subdue" and dominion."

- ❧ Humankind are made in the "image of God" with the responsibility to be stewards and a blessing to all creation (the environment).

- ❧ As stewards, people recognize the right of all species to exist. God's delight in God's creation is not limited to *Homo sapiens* (Gen. 1:31).

- ❧ Rather than being in the forefront on environmental issues, Christians are usually complacent or defensive. The latter may be true because voices both within and outside of the church are leery or hostile towards Christian environmentalists.

- ❧ Some Christians with anti-environmental beliefs scorn the perspectives of "secular humanists." These brethren do not realize that in doing so they place themselves in the same camp as the secularists who separate God from God's creation.

- ❧ Christian environmentalists are Trinitarian. Central to the convictions of Trinitarian Christians is faith in Christ Jesus, who is reconciling the world to himself (Col. 1:20).

While there are several common themes, the contributions of each author are distinctly his or her own. Complementary subjects are discussed that vary according to the training and interests of the authors. The Reverend Peter Harris ("Environmental Concern Calls for Repentance and Holiness") is a practicing ecologist. Harris is deeply troubled by the anti-environmental attitudes found in much of Western culture, including the church. Repentance is called for, as many Western Christians live their lives as practicing atheists, corrupted by their cultures. The Enlightenment divorced God from creation and emphasized the role of individuals outside of community. Harris stresses that we live in a created world that's intensely relational— we to God and humankind to the rest of creation. This is a world where the spiritual transcends the material, and one in which God is interested in us now as well as throughout eternity. Christian stewardship is a sign of hope in a desperate world looking for meaning.

We needn't read much of Dr. R. J. Berry's chapter ("One Lord, One World: The Evangelism of Environmental Care") to realize that we're in the presence of a scientist. Berry cautions us to examine Scripture with God's intentions in mind, rather than our own biases. In so doing we'll see in the "creation ordinances" how humankind were to behave and interact towards the rest of creation. Words such as "dominion" and "subdue" are best rendered as caring support or stewardship. The basis for this interpretation is that people are made in the image of God. This—not genes or DNA—is what makes humans different from the rest of creation. Being made in the image of God implies a relational existence to God, to each other and to creation. Further, humankind should be a blessing to all creation and lead creation in praises to God. To be a steward and a blessing, the first step is to become environmentally literate. Then we may wish to join and/ or work with environmental organizations.

Ethicist Dr. Michael Northcott follows with a chapter entitled "Ecology and Christian Ethics." Northcott begins with a personal testimony of how consumer desires in affluent countries contribute to tropical forest desecration in developing nations. The author then asks who is to blame for today's ecological crisis. Many reasons may be found, and not only a simplistic reductionist approach that faults a religious tradition (i.e., Western Christianity). Northcott critiques different Christian approaches to the environment, including the limitations of process theology and deep ecology. His own view is that "humanity is bequeathed the earth not as a possession but as a gift." This gift is part of a covenant with God that includes the bios and connects human injustice with ecological disasters—circumstances as true today as they were in the times of Israelite prophets of old. Northcott goes further and develops the thesis that the Resurrection is the most fundamental ecological concept. Here, God reconciles all of creation to himself through redemption from sin. Northcott closes with several elements that constitute "Christian environmental ethics." The first and most essential is a confession of our individual and collective despoliation of creation. As healing and redemption for individuals come through Christ Jesus, so, too, come healing and restoration of the land.

Theologian Dr. Anne Clifford ("From Ecological Lament to a Sustainable *Oikos*") first relates God's care for the environment through Jeremiah's lament for the land. God's concern for the environment is as true today as it was in the past. If we desire an *oikos* (earthly home) that will adequately support humankind, then Christians must advocate for a sustainable environment. Out of the many challenges to a healthy *oikos*, Clifford discusses two. The first is that when it comes to the environment, Christianity is often seen as an anthropocentric religion. Clifford dismisses this in a discourse that ranges from the Creation stories in Genesis to Bacon and back to Genesis in the covenant between God and Noah, his descendants and every living creature. Clifford's second point is that redemption through Christ Jesus does not exclude, but rather emphasizes, humankind's obligations to the land (bios). God's redemption is cosmic and includes all of creation.

The editor, Don Brandt ("Stealing Creation's Blessings") echoes several of the themes discussed by the other authors. Brandt emphasizes that the normative role of Christians is as Trinitarian environmentalists. That role is played out in everyday life as God's stewards and caregivers of creation. While the ecological conscience of Christianity has often been marginalized within the church, it has always existed. Today, the care of creation is recognized as a Christian responsibility by Roman Catholic, Orthodox and most evangelical and conciliar Protestant churches (see Appendix).

---

[1]  In his chapter, Berry makes a point that all "people of the Book," Christians, Jews, and Muslims, should be included as prime environmental caretakers.

[2]  This widely used definition is from a report by the World Commission on Environment and Development, *Our Common Future* (New York: Oxford University Press, 1987). See the "Conclusions" section in Clifford's chapter for comments on this definition.

[3]  The term "transformational development" is used in World Vision to indicate that attitudes toward the Creator and creation usually need to change before sustainable development is achieved.

**1**

⤫

# Environmental Concern Calls for Repentance and Holiness

*Peter Harris*

## Introduction

In coming to terms with the imperative for environmental stewardship in recent decades, we must recognize that the church is rooted in cultures that are enormously numerous and varied. Inevitably, individuals and religious communities around the world approach the issue of caring for the environment in various ways. Christian thinkers and leaders, including environmentalists and other professionals, have concerns that are extremely diverse. The issues that confront a park ranger in East Africa are a planet away from those that confront a geneticist working in a lab in France. The concerns of a pastor in Guinea Bissau, where deforestation and collapse of biodiversity spell starvation and disease for parishioners, have little in common with those of a minister in a wealthy suburb in the Western world, where materialism has all but overwhelmed any desire for sacrificial living.

Regardless of the specific situation, though, an authentic Christian response is always going to be formed by changing circumstances in relationship to God. Consequently, some common themes may be found. But before looking at the rich heritage of Christian belief and practice that is now leading to a renewed concern for the creation in many places, we must take an honest look at the context out of which it is developing. In doing so I will only consider the context of the Western world. This is partly because I know the Western world best, and partly because it is Western culture—whether on home ground or in some exported form—that is driving much of the current environmental degradation taking place worldwide.

If we must begin on a somber note, we do well to remember the words of Rowland Moss, who as a human ecologist and a Christian was one of the first in recent times to recall the church to its environmental conscience:

> Is the environmental challenge to . . . the church but a part of the total moral challenge which is greater than the sum of its individual components . . . Is the whole situation much more serious than we think (not for scientific or economic reasons, but for moral reasons) in that we are already under God's judgement?"[1]

Rowland Moss believed that our first response to the current environmental crisis should be repentance. If we accept his moral argument, it then gives us a way forward to consider the often-overlooked role of belief itself in the environmental debate. Then the remarkable resources that belief in God brings us will help us as we reconsider and renew our relationship to the world around us.

## Sleeping with the enemy

Sadly, church history reveals time and again the uneasy compromises Christians make with their times. Examples appear in a sobering parade, from the superstitious practices of the twelfth century church, to the defense of slavery and racism on the part of churches in the eighteenth century, to the spiritual aberrations that condone nationalism in recent history. Our times and cultures tend to cling to us very closely. Christians know that it is only as the Holy Spirit renews our minds, as God reveals himself in Christ and in the Scriptures, that we are delivered from being pressed into the distortions of the cultures in which we live. Many contemporary Western societies are determined to edit out from their experience of the world any recognition of God as Creator. Christians often find it difficult not to continue living and thinking as they did before they came to know Christ. It can be very tempting to reach an accommodation with society so that a Christian profession of faith is so weakened that it becomes "practical atheism." While calling themselves Christians, people may live as if there were no God in terms of choices or preferences they make and in the general manner in which they live their lives.

On the other hand, Western societies share a powerful Christian heritage. Christians have become accustomed to the idea that being a good Christian is the same as being a good Dutch, French, American or British person. But Christian culture, if it ever truly existed in the past, is far from a reality now. The time has long gone since Christian thinking exercised a wide influence on policy makers. Rather, in seeking to reconcile social consensus with Christian beliefs, on many issues the Christian is more akin to a person with a foot on two ice floes that are drifting steadily apart. Although a choice must be made, the idea of critical evaluation is unfamiliar to Western Christians. We are unaccustomed to thinking that Christians cannot simply belong to our culture and to our church by the same set of decisions. There is a further challenge; namely, to take seriously the implications of what we believe about God and the world around us, and to apply that to a number of critical issues. In reality, it has never been possible to be unthinking members of cultures that deny the lordship of Christ, even in times of consensus between church and state. Today, more than ever, nearly impassable chasms exist between the paths of journeying with Christ and the roads of errant direction espoused by secular societies.

## Humanism

A prime example of Christian beliefs corrupted by cultural values occurs in the arena of environmental issues and concerns. Worldwide, we are faced with an unprecedented and extremely rapid loss of biodiversity[2] and the collapse of many ecosystems that have sustained life on earth. Straightforward Christian convictions make it impossible for the church simply to endorse the choices of a destructive culture. But in understanding the roots of the problem we will need to look deep into the soil of Western beliefs. As we become aware of how those beliefs have taken shape in the development of industrial and post-industrial societies, we have to recognize one of the more disastrous accommodations of the Christian church. This is a well-documented cohabitation with the individualistic humanism that flourished in Europe and elsewhere in the Western world for the last four or five centuries.

Where this cohabitation became established, individualism had the effect of reducing Christian thinking to a merely personal agenda, excluded from guiding social or environmental relationships. It led to a conception of the creation as merely the raw material for economic growth. In so far as societies that set off down this road were posited on Christian assumptions, we bear some responsibility for the outcome. In hindsight we can see clearly that such a view of creation is far from being coherently Christian. A more biblical reading of basic beliefs will help us to discover the road back to a lifestyle that is more consistent with our core beliefs and that will lead inevitably to a proper care of creation. Before we can go on to consider what those beliefs are, and where the implications of them might lead us, we need briefly to look at another authentically Christian conviction that is at variance with Western secular assumptions about environmental issues; namely, that belief itself is relevant.

## The centrality of belief

For Christians and many others, belief itself is central to life. Belief cannot be seen merely as an optional extra for those who like to consider themselves religious. What we believe the world to be will profoundly affect how we treat the cosmos and how we live in it. James Houston, a former Oxford University geographer, now theology professor, wrote:

> The world we see is the mirror image of our hearts. We perceive reality as we conceive it to be. If we have given up the hope of finding meaning in our lives, then we see the world as a desert place, a threatening wilderness. If, however, we have hope in God the Creator of all things, then we can, and shall, see it very differently.[3]

To adapt Houston's words, if today we are making a desert of the world we live in, it is because of the prevailing beliefs of our times. Even if some value is given to the concept of biodiversity, the usual argument for its preservation begins from our need as a human species for access to the globe's flora and fauna riches. The assumption is that biodiversity must be maintained primarily for the well-being of humans. In the disappearing flora of the rainforest, it is argued, may lie many remedies for human diseases. In this anthropocentric worldview, the significance of biodiversity is merely that of a belief—and a remarkably inconsistent and untenable one at that. Yet it is undeniable that such human-centered thoughts are pivotal in how we make our decisions.

Even so, it is extremely rare for those most concerned about the well-being of biodiversity in the major world institutions to give more than token consideration to this "belief basis," despite its fundamental role in determining how we treat the world around us. There are some notable exceptions, but the debate tends to focus far more on the "how" questions than the "why" questions, which are of equal and prior importance. While obvious beliefs undergird all our choices, in Western society at least, there is a constant and strenuous attempt to edit out the discussion of belief from the business of government and public life generally.

## A starting point for belief

Part of the contemporary Christian re-appraisal of our environmental responsibility stems from the widespread discovery that authentic Christian belief—revealed in Scripture and known through Christ the Creator—gives the most compelling basis for a way of life that does justice to creation itself. Having acknowledged that this belief has frequently been put aside in practice, and having argued for the relevance of belief itself, we can now consider how a true relationship to the Creator should guide the Christian church in renewing its relationship to creation.

### A created world

From beginning to end, the Bible affirms that God is the Creator. This, and not the human condition, is the true starting point for both understanding and caring for the world around us. We discover who we are and what the world is in relationship to God, rather than the other way around. Nevertheless, this understanding of God is at odds with the conviction that achieving human happiness is at the heart of the human enterprise, and thus of our environmental relationships. Assertions that belief in the Creator is not primarily an affirmation about humankind come as a big shock to most people, including the majority of Christians.

We have to recognize that we are not saying, "I believe in God who made me" but, as the Apostles' Creed states, "I believe in God, maker of heaven and earth." We are put quite literally in our place if we begin by recognizing who God is. Such recognition stands in direct contradiction to our popular assumption that we are at the centre of all things. We need to notice that it is not simply Christians who can be tempted into this humanist heresy. Even those in the secular world who are moved to re-establish the importance of the non-human environment can easily stray into overstating the final significance of people either as managers or destroyers. In Christian thinking we understand that we are, first, in relationship with God and then in relationship with creation.

## Relationship with God

This relationship with God and creation goes beyond a simple affirmation and begins to seem more important. In Christian thinking, the God who made us has revealed himself to be personal. God is calling us to relationship with himself and thereby to the renewal of all our relationships. A simple glance at the Creation shows how intensely relational the whole cosmos is. Ecology is the study of those relationships in the widest sense. Not only we as humans but also creation itself can have a relationship with our Creator, if all around us is God's handiwork. And by virtue of being part of that same creation, albeit with a unique capacity for knowing God, we are in an inevitable relationship with creation. That relationship can be lived out either faithfully, in the context of our relationship with God, or abusively, without reference to him but merely according to our own wishes and needs.

The pervasive contemporary idea is that most of life is "religiously neutral," even if it is admitted that there are some questions that may be considered religious. In contrast, a biblical perspective is that "for in him [Jesus] all things in heaven and on earth were created, things visible and invisible" (Col. 1:16a, NRSV). Here is a theology that frames everything we know in time and space as God's artifacts, the work of his hands. By contrast, Western society has offered the church a kind of golden cage where it can be left in peace to pursue the "religious life." The church is permitted to comment quietly on such issues as how to run religious services, or how to make new translations of the Bible, or maybe on a few moral issues such as abortion. This narrow path is often reckoned to be the proper sphere for believers, and it is made clear in many ways that they must never venture into discussions of economics, the environment, the arts or public health. In turn, Christians have too often been happy enough to accept the reduced space that is offered to them for the expression of their belief, and consign the rest of life to unbelieving pseudo-neutrality.

Instead, we need to take into our minds and hearts the conviction that our relationship with God, whether lived as rebellion or belief, is fundamental to the shape taken up by all that we see around us. Creation itself becomes more fully understood as we ourselves enter more completely into a renewed relationship with God. Our reference points and even our language begin to change. Thus we live in "creation," not "the environment" or "nature," and we ourselves are in a different relationship with those around us, sharing a common created humanity, and not an identity which is fractured by nationality or even by creed.

The apostle Paul, whose thinking was profoundly shaped by his understanding of God the Creator, viewed people as "God's offspring." Paul happily quoted from a contemporary pagan poet: "'In him we live and move and have our being'; as even some of your own poets have said, 'For we too are his offspring'" (Acts 17:28 NRSV). Disintegration of relationships of all kinds is now deeply embedded in our Western societies. It is clear in the destruction of community and personal relationships at all levels. It also extends to the destruction of meaningful relationships with anything we consume. Food appears we know not from where, our machines are often produced in appalling conditions in parts of the world we never see and the waste we produce is "spirited away"—but as Christian environmentalist Loren Wilkinson asks, where in creation is "away"?

In contrast to this individualistic and isolated existence, Christians affirm their belief in all kinds of interdependencies and relationships simply by affirming their belief in the Creator. Of course, this insight is entirely confirmed by the observations of any working ecologist or biological scientist, as created reality itself bears witness to the character of its Creator.

## One created reality, visible and invisible

Belief in God as Creator brings a further implication that we must come to terms with; namely, that God made heaven and earth. For the Christian there is one created reality, visible and invisible, and any idea that only the material, or visible, is real is far from Christian understanding. Equally unique to Christian thinking is an idea, often part of popular Christian imagining, that somehow the word "spiritual" refers to the non-material. Biblically, what is "spiritual" is simply whatever is the fruit or work of the Spirit of God—who as the Psalmist understood is the Creator of, among other things, the material: "O LORD, how manifold are your works! . . . When you send forth your Spirit, they are created" (Ps. 104:24, 30, NRSV).

If we believe that "spiritual" means non-material, then of course it is reasonable to suppose that much of what we do, and all that exists around us, is of supreme indifference to God. But if we believe as the Bible insists—that matter itself is God's handiwork and is created for his glory and praise—then our relationship to it is firmly in the realm of the spiritual. God is interested in how we make our money, what we eat, what we watch on television and how we treat the environment. These are "spiritual" issues. They are spiritual not least because as the Psalms also explain: "The earth is the LORD's, and everything in it" (Ps. 24:1a, NL).

### *Where or what is heaven?*

There isn't space to do more than briefly refer to one final but important point about heaven. One of the most significant changes to take place in Christian thinking in recent years would seem to be a rediscovery of the biblical understanding of eternity. The evidence in Scripture is that creation itself is waiting, with us, for renewal, and for redemption. In the words of John Stott,

> It would not be wise to speculate how the biblical and scientific accounts of reality correspond. The general promise of the renovation and transformation of nature is plain . . . God's material creation will be redeemed and glorified.[4]

Immediately this rescues creation from any suggestion of irrelevance and the church from the heresy that we sometimes hear expressed: "Ah well. It is all going to burn up in judgment, so why should it matter what happens to it meanwhile?" It is a good thing that we don't treat our bodies, the part of the physical creation with which we are most intimately concerned, in the same reckless fashion!

## The gospel is for the whole earth

But if there is an environmental implication to be drawn from the Christian understanding of heaven, there is a further important lesson to draw from the Christian understanding of life on earth; namely, that the Christian gospel is for the whole creation. Once again it is the apostle Paul who makes this clear: "that the creation itself will be liberated from its bondage to decay and brought into the glorious freedom of the children of God." (Rom. 8:21, NIV). In A Rocha,[5] we are trying to put into practice that message of hope as a sign of the coming kingdom—to make a difference to disappearing habitats and polluted wasteland, as well as to the injustice suffered by the human communities that depend upon them for their life. Notice how Paul in his Acts 17 sermon insists that the fact that God is Creator automatically means the gospel is relevant to all people as well. Paul's words "we are God's offspring" (Acts 17:29, NIV) confer absolute equality among all people as we are created "in the image of God" (Gen. 1:27, NIV).

This insistence on human community is a vital corrective for Western societies that marginalize the elderly in their adoration of youth, that are increasingly drawn to nationalism, and that are built on the practice of racism in their overseas policies. As a European, I believe it is the loss of our belief in God the Creator in post-Christian Europe that is the root cause of a multitude of ills in our human relationships. Ironically, an attempt to overcome the problem by an idealization of the human, which in the media takes a highly selective and unreal view of what the term means, only leads to further contempt for real people. Once again, people find their dignity and meaning only in relation to their loving Creator God, and not in competition with the super-rich and super-beautiful.

### What is our response?

A response is required from us today, whoever we are and whatever we believe. If we don't believe in the Creator then we must look hard at the implications of the other bleak possibilities. Do we really live consistently with the idea that we are merely an accident and the product of unknown forces? Does environmental concern itself make any sense if we are simply part of a blindly determined process and an unseeingly competitive struggle for survival? But if we affirm that we are living in a created world, are we ready to respond to its constant prompting to reach out for the Creator? And if we actively acknowledge this, are we also ready to treat the world around us as important to God, and not merely there to meet our needs?

So, if we are Christians, we face the big challenge of living in the world as creation. We face the challenge of allowing Jesus Christ to redeem all of our relationships, in all their brokenness and complexity, so that we begin to acknowledge him as the Lord of Creation and not merely a personal Savior in some reduced private space. Then we can recognize with Paul that, "In him we live and move and have our being" (Acts 17:28a, NRSV). If we do not, we face the far more serious challenge of establishing any coherent or widely compelling grounds for environmental responsibility that can begin to reverse the almost universally damaging trends that are now established on every continent.

---

[1]  Rowland Moss, *The Earth in Our Hands* (Leicester, UK: Intervarsity Press,1982).

[2]  The World Conservation Union (IUCN), *The Red Data Book, 2001* (Cambridge, UK: IUCN), 2001).

[3]  James Houston, *I Believe in the Creator* (Hong Kong: Regent Publishing, 2000).

[4]  John Stott, *The Message of Romans* (Leicester, UK: Inter-Varsity Press), 1994, 240.

[5]  "An international conservation organization working to show God's love for all creation."

**2**

🖎

# One Lord, One World: The Evangelism of Environmental Care

*R. J. Berry*

## Introduction

It is trite to say, but the world is getting smaller—not in the sense that bits are falling off nor that the earth is physically shrinking. It is getting smaller because we are now only too aware that almost everything on the globe is within touching distance—a mere day's flight or even a microsecond away by TV or e-mail. The finiteness (and fragility) of this planet became frighteningly obvious on September 11, 2001. How different this is to the world of our grandparents, who used to plan for months if they intended to travel to what they called the "uttermost parts of the earth," while *their* grandparents ventured beyond "civilization" only at certain risk to life and limb. Three centuries ago, the edges of world maps were decorated with monsters and labeled "dragons be here." The seventeenth century Europeans who sought their "promised land" in North America were frightened by the apparently never-ending forests. Frequently they commented in letters and diaries about how wonderful it was to cut down trees so that they could see the stars or, even better, a neighbour and know that they were not completely alone.

Our theology has not kept step with these changes in our perception of the world. For ancient Israel, the sea represented fearsome disorder, the heavens were a firm ceiling above the earth and it was ridiculous to think of a moving planet. We now know differently. We are not disturbed by the knowledge that the Earth moves around the

Sun; we do not expect to "fall off" the world when we sail beyond the horizon. But we still tend to read the Bible as "flat earthers," looking for scientific teaching about the physical world. We have not yet caught up with Galileo who wrote nearly four centuries ago that "the Bible teaches us how to go to heaven, not how the heavens go."

The Bible cannot teach science, because God's Word has to be understandable by men and women throughout time. If it were written with the vocabulary and scientific jargon of a twenty-first century textbook, the Bible would have been unintelligible to our forebears of only a few generations ago. God has given us his Word in a *non*-scientific form—not *un*scientific, implying subjectivity and error—in a language that is eternally comprehensible. When the Psalmist wrote, "The world is firmly established, it will not be moved" (Ps. 96:10, NASB), he was not pronouncing on cosmology, but proclaiming the absolute control and certainty of God's rule, as the rest of the verse makes clear. When he said "the sun . . . rises at one end of the heavens, and follows its course to the other" (Ps.19:5,6, NL), he was not describing the size and limits of heaven or insisting on a static earth, but putting into words an everyday observation. God's Word is flawless (Ps.18:30), eternal (Ps. 119:89 ) and truly "useful for teaching, rebuking, correcting and training in righteousness" (2 Tim. 3:16, NIV), but we have to come to it afresh in every generation to seek its relevance and implications for us.

This brings us back to our shrinking world. If we look at our surroundings in the same way as the original writers of the Scriptures or even through the eyes of the great seventeenth century Reformers, we are likely to be led badly astray. This becomes important in practice when we start using non-biblical but common modern words like "environment." Gro Harlem Brundtland, director of the World Health Organization and a former prime minister of Norway wrote this in her foreword to the United Nations Commission on Environment and Development that she chaired:

> The environment does not exist as a sphere separate from human actions, ambitions and needs. Attempts to defend it in isolation have given the word "environment" a connotation of naivety in some political circles. The word "development" has also been narrowed by some into a very limited focus, along the lines of "what poor nations should do to become richer," and thus again is automatically dismissed by many in the international arena as

being a concern of specialists, of those involved in questions of "development assistance." But the "environment" is where we all live; "development" is what we all do in attempting to improve our lot within that abode. The two are inseparable.[1]

What would the Puritans have made of this? Is it defensible in this newly shrunken world for our leaders to abstain from properly caring for our common environment lest they alienate their supporters? Can we learn anything from the Bible about our proper attitudes to "the environment" or "development"?

## The Bible is relevant

The word used in Scripture for what we call "environment" or "nature" is "creation"—what we know to be God's work, not through reason or science but through God's revelation in the Bible: "In the beginning God created the heavens and the earth" (Gen. 1:1, NASB); "*By faith* we understand that the universe was formed at God's command" (Heb. 11:3, NL).

Scripture tells us in many places and ways what our relationship to creation is or should be. The very first command given to our original parents is not about love or social organization or work or sexual relationships: "Fill the earth, and subdue it; and rule ["have dominion"] over the fish of the sea and over the birds of the sky and over every living thing that moves on the earth" (Gen. 1:28, NASB). The commands in the first few chapters of Genesis are sometimes called "creation ordinances." They describe how God's creatures should function, in the same way as an instruction manual shows how a piece of machinery should be used so as to work efficiently and without damage. And just as a manual usually has a list of warnings, so the Bible sets out a series of laws to help us monitor and recognize our design features. The creation ordinances are positive guidelines; they are perhaps more familiar when expressed negatively in Exodus and Leviticus as a series of prohibitions ("You shall not . . .").

### Dominion as care

The command to humans to "fill the earth" repeats words addressed also to animals in Genesis 1:22; it is a general biological instruction and not one limited to humans.[2] The instruction to "subdue and have dominion" (the words are probably equivalents) has caused long-standing problems. University of Pennsylvania professor Ian McHarg described it as

> one text of compounded horror which will guarantee the relationship of man to nature can only be destruction which will atrophy any creative skill . . . [and] explain all of the despoliation accomplished by western man . . . The Genesis story in its insistence upon domination and subjugation of nature encourages the most exploitative and destructive instincts in man, rather than those that are deferential and creative . . . God's affirmation about man's domination was a declaration of war on nature.[3]

McHarg's denunciation of human attitudes reflects a very common misunderstanding and shows why many environmentalists regard Christians as at best unhelpful and more probably antagonistic towards creation care. Yet McHarg and those who think like him have misread the Bible. Culpably, Christians have failed in allowing this misreading to go unchallenged. The difficulty is that the word translated "dominion" is one that describes kingly, authoritarian rule. However, we need to remember the Hebrew ideal of kingship. Although most of the kings of Israel and Judah were despotic tyrants, God's model was that of a servant-king typified by David or Solomon or our Lord himself. Our dominion is properly—and scripturally—exercised as a caring support for creation, not as an unfettered right to plunder (Ps. 72).

This interpretation is underlined when we follow the invariant rule of Bible understanding and take into account the context of the command to "have dominion." In the same verse (Gen.1:26) that we are given dominion, we are told God made people "in his image." Our views about evolution and human origins are irrelevant. The Bible is explicit in explaining that "image" is the element that distinguishes us from every other animal. We share many traits with other animals, including having more that 98 percent of the same genes as chimpanzees. What makes us human is not genetic or anatomical; we are human because of God's image in us. Theologians have filled volumes inquiring about what is meant by "God's image," but there is general agreement that it implies relationship with God, with each other and, less obviously, with God's work in creation. Furthermore, "*God's* image" implies reliability and hence responsibility. The God of the Bible is utterly trustworthy. He has shown his steadfastness to people throughout history despite their (which includes our) faithlessness. It follows that God expects us to be faithful—to him and to the tasks he has given us.

Genesis 1 tells us that God created the world and everything in it, and that he has committed it to us to look after for him. Genesis 2 describes Adam as a gardener, appointed to tend and care for the place where he has been put. The Bible is unequivocal in stating that the world belongs to God (Ps. 24:1) but that he has delegated its care and protection to us (Ps.115:16). We have no license to plunder creation, but we have a charge to care for the cosmos that God loves (John 3:16) and that Christ has redeemed (Col.1:20). We are not asked to look after a world that is only a "thing"; we are required to be God's agents in managing a world that he created, redeemed and sustains.

### Biblical misunderstandings

This biblical responsibility is commonly misunderstood and ignored. Christians are blamed for "de-sanctifying" nature, degrading an allegedly sacred earth into mere matter. During the 1960s in particular, many people turned to the pantheism and mysticism of Eastern religions in the belief that they somehow offered salvation for the environment. This proved vain. Jacques Delors, a former president of the European Commission commented:

> The Oriental religions have failed to prevent to any marked degree the appropriation of the natural environment by technical means . . . Despite different traditions, the right to use or exploit nature seems to have found in industrialised countries the same favour, the same freedom to develop, the same economic justification."[4]

Japan and the former Eastern Bloc countries are among the most polluted in the world, although they owe comparatively little to Christianity.

Another attempt to find a truly "green religion" has been through the New Age, a seductive but amorphous creed marked by the world moving from astrological rule by Pisces (the symbol of Christianity) to the barrier-less Aquarius (the water carrier). The New Age blends all things together so that we are merged into a massive cosmic unity, with no distinction between spirit and body, or between god(s) and humanity. New Age adherents often claim credibility for their beliefs through Jim Lovelock's Gaia Hypothesis that postulates complex feedback between life and atmosphere—that the cosmos functions as a vast single organism (called *gaia* after the Greek earth goddess). As science, this hypothesis has been valuable in stimulating research, but we must be very wary of extrapolating beliefs (or even facts) about the material world into the metaphysical realm. The assumption that the earth and life form a single unit is more mythical than realistic. It should not be used to support religious beliefs.

## God and the environment

We can state the biblical basis for creation care in four propositions:

**1. God works in the world.** The best way to understand the connection between divine and natural causes is "complementarity." For example, a painting can be described "scientifically" in terms of the nature and distribution of chemicals on a two-dimensional surface, but it can also be described in terms of the plan and intention of the artist. We can have two (or more) descriptions of the same object that do not overlap or contradict in any way; we call them complementary. In the same way, God can be understood by faith (Heb. 11:3) as working in the world without conflicting or diminishing any scientific (or rational) knowledge we may have of the same event.

**2. God is separate from creation.** The world is not an extension or emanation from God. Although it is not explicitly stated in Scripture, from early Christian times it has been believed that God created "from nothing." If God had used existing material, this would mean that he was dependent on something outside himself, and hence would not be "before all things." As humans, we are a part of nature but we are unique in being made in God's image and responsible to him.

**3. Christ has redeemed *all things* by his death on the cross** (Col. 1:20). Christians often base their teaching about creation care solely on God's creating work as described in the early chapters of Genesis. This is wrong; it ignores other significant passages in the Bible. The Christian faith is Trinitarian, not Unitarian. As the Father creates, so the Spirit upholds that which Christ has redeemed.

**4. Our response is (or should be) stewardship,** and stewardship means active management—not merely conscientious preservation, however devoted. The man who concentrated on guarding his talent so as to hand it safely and unchanged back to his master was the man who was roundly condemned as a "useless servant" (Matt. 25:30, NL).

The relationships made possible by God's image in us involve continuing and diligent care. We have an ongoing task. The statement in
Genesis 2:2,3 that God had "finished" his work refers to a pattern of
rest for the seventh day in every week, not to a situation where he
ended his work once and forever. After all, the Creator we worship
never rests (Ps.121:4, John 5:17).

Stewardship is only part of our proper response to God. We may
be good or bad stewards. We may get occasional twinges of conscience
about how faithful or efficient we are in our obedience. But underly-
ing any work we do in this life should be the recognition that we are
mere strangers and pilgrims; that we are living in a world that belongs to God. When we change the landscape, use "resources" (such
as farm products or fossil fuel) or even merely travel along a highway,
we are in territory that is not our own. Our God is a jealous God. We
cannot escape his anger if we ignore or spurn him. Bad behavior is
one thing; disregarding God is much more blameworthy in his sight.
The Ten Commandments begin, "I am the Lord your God . . . You shall
have no other gods before [or as rivals to] me" (Exod. 20:2,3, NRSV).
The Old Testament focuses on a God who is all-knowing, and abso-
lutely holy. Our ultimate offense is to reject the Creator and sustainer
of this world.

## The Fall

What about the Fall? Although we were given a charge at the beginning to be stewards, we now live in a fallen world. What does that
mean? The man (Adam) was told in Genesis 3:17 that because of his
disobedience the earth would be cursed. In other words, there would
be a significant change in creation from the state when God saw all
that he had made and declared that it was "good . . . very good."

We need to be clear that when God called his work "good," he
was speaking from his viewpoint, not ours. Too often, we project
our own ideas into imagining life in the Garden of Eden. In particular, we have to recognize that there was certainly death in the world
from the earliest days of biological life, not least because God explicitly gave plants to animals for food (before the Fall), and plant
death is as much death as is animal death. Indeed, there were many,
many generations of animal death before humans came on the
scene. We are wrong to assume that there was no biological death
or disease before the Fall. We must be very wary about adding our

personal suppositions to Scripture.

The "death" that entered the world with Adam (Rom. 5:12, 1 Cor. 15:21) was primarily separation from God, the life-giver and source of that which makes us truly human. Our first parents "died" the day they sinned. They were banished from God's presence (i.e., evicted from Eden), but they survived for years and had all their children outside Eden. *We* were dead. It is only through Christ's redeeming death that we are raised to life, "born anew" by being reunited with God (Eph. 2:1, 5).

The key to understanding the Fall is in the New Testament. Interestingly, neither the rabbinical nor the Jewish apocalyptic tradition has any concern about a fall into sin. Judaism emphasizes individual responsibility for our failings, not some intrinsic curse upon creation. The Fall is a Christian doctrine and this makes it particularly important to interpret the New Testament references carefully.

The most important passage for understanding the Fall is Romans 8:19–22 (REB), where we are told that the created universe

> was made [became] subject to frustration . . . yet with the hope that the universe itself is to be freed from the shackles of mortality . . . Up to the present, the whole created universe in all its parts groans as if in the pangs of childbirth.

This is a difficult passage, and most commentaries offer little help. One theologian who comes to grips with it is Charles Cranfield in a magnificent *reductio ad absurdum* argument. He asks:

> What sense can there be in saying that the sub-human creation—the Jungfrau, for example, or the Matterhorn or the planet Venus—suffers frustration by being prevented from properly fulfilling the purpose of existence? The answer must surely be that the whole magnificent theatre of the universe, together with all its splendid properties, and all the varied chorus of sub-human life created for God's glory is cheated of its fulfilment so long as man, the chief actor in the great drama of God's praise, fails to contribute his rational part. The Jungfrau and the Matterhorn and the planet Venus and all living things too, man alone excepted, do indeed glorify God in their own ways, but since their praise is destined to be not a collection of individual offerings but part of a magnifi-

cent whole, the united praise of the whole creation, they are prevented from being fully that for which they were created to be, so long as man's part is missing, just as all the other players in a concerto would be frustrated of their purpose if the soloist were to fail to play his part.[5]

Other expositors have made the same point. For example, Derek Kidner describes the cacophony produced by the Fall. Commenting on Genesis 3, he writes:

> Leaderless, the choir of creation can only grind in discord. It seems from Romans 8:19–23 and from what is known of the pre-human world, that there was a state of travail in nature from the first, which man was empowered to "subdue," until he relapsed into disorder himself.[6]

God made and appointed us to be stewards, managers, agents and caretakers. By our disobedience, we have not only removed ourselves from fellowship with and support by our Creator, but also condemned all the rest of creation into a state of disorder and consequent inability to fulfill its role (Ps. 19, Ps. 148). The French theologian Henri Blocher argues that

> if man obeys God, he would be the means of blessing the earth, but in his insatiable greed . . . and in his short-sighted selfishness, he pollutes and destroys it. He turns a garden into a desert (cf. Rev. 11:18 [7]). That is the main thrust of the curse of Genesis 3.[8]

From beginning to end, the Bible speaks of our links with nature. Sometimes we are given direct commands, as when we are told to "have dominion." On other occasions, the instructions are implicit: the perils of a journey, the care needed for a farm or a flock of animals, our relationships to wild animals or fierce weather. The Bible repeatedly connects human behavior with responsibility to the environment. Sin led to the Flood and also to drought (Lev. 26, Deut. 28); the food laws regulate hunting; a very positive attitude towards nature is set out in Job, Proverbs and the Song of Songs; and so on. Throughout, we are described as interacting with creation, a part of it as well as *apart from* it. But also throughout is a parallel theme: that this is God's world; that he made a covenant with us that he will not break; and that both creation and we as people were reconciled to God by Christ's death on the cross.

God's message comes to us in various guises. It is sometimes described as being in two books, one of words and the other of works. God is faithful, despite our rebellion and faithlessness, and calls us to be obedient and to respond to him. Only then will we enter into the inheritance prepared for us. Only then will the universe no longer groan in frustration.

## God's unfailing providence

So far, so good (or bad). We live in a disordered world, longing and praying for the time when we will join with all creation in the "glorious liberty of the children of God" (Rom. 8:21, NKJV). But we can become too depressed. Environmentalists delight in telling us that the worst pest and greatest polluter is humankind. A common cry is for us to escape from the urban squalor into the unsullied wilderness and so find God in the world as God made it. Is this right? Are the effects of the Fall only obvious where humans have fouled their nests and their surroundings?

The answer to these questions is an emphatic "no." We are stewards appointed to weed and prune God's garden. The situation becomes even clearer when we consider the boundary where science and religion meet. The medieval worldview was built on Greek ideas of a perfect and unchanging order in nature. The natural world as a whole was often compared to an organism—a macrocosm under stress and with inter-connections and disorders just like any individual animal. Paul uses this analogy when he compares the church to a living body (e.g., 1 Cor. 12:12–27). Now one of the characteristics of a living body is that it has considerable powers of self-regulation. A fever subsides; a wound heals; tiredness is overcome. If that is so for an individual, is it true also of nature?

In fact, it is not. It is inaccurate to speak of "balance" or "equilibrium" in nature, although it is easy to be misled. If we burn a forest, in time it will regenerate. If we abandon a cultivated area, it will revert to a wild state. But the error of assuming some sort of natural equilibrium persists, helped by some environmental philosophers who are fond of envisaging the world as akin to a body, frequently using the Gaia Hypothesis as scientific support for their ideas. This confusion is increased by a hangover from early twentieth century ecologists who overemphasized the connectedness of living systems, particularly by assuming that different habitats would progress inevitably to a stable "climax" characteristic of a particular area (such as temperate woodland, sheltered rocky shores and high altitude meadows). Not surprisingly, politicians picked up the notion of "balance" in nature as part of their rhetoric, with the resulting assumption that anything upsetting the balance is demonized. While it is obviously true that ecological communities often change (or "mature") with time, it is false to think of the whole of nature as an independent and harmonious "web."[9] A disturbed area is likely to be re-colonized from its near surroundings and hence will appear to "recover," but we must be clear that this is very different from an automatic readjustment to a pre-existing ideal.

Seen from the viewpoint of an individual, the world is so big that it seems impossible that we can change it in any way, for good or ill. This is a dangerous error. Our world is not a living body with an ability to recover from assaults in the way we do, but a massive edifice with inertia to change because of its size and with no automatic mechanism for recovery from damage. *We* are its divinely appointed guardians. If we find that we are harming our environment, we have to examine whether our behavior is so serious that it is affecting our own survival and quality of life, never mind that of our children and grandchildren. If we are Christians, we also have to ask ourselves a more searching question: "Can our actions be regarded as 'tending the garden' that God has entrusted to us?" Are we ignoring the Master Gardener, which is a sin just as much as greed or adultery? If we are "people of the Book" (Jews, Christians or Muslims) we should be leading in creation care. Christians ought to be in the forefront here, because we can testify that the God who creates and sustains has also redeemed. We have good news: creation care is an evangelistic opportunity as well as a religious duty.

## One Lord, one world

The biblical doctrine of the environment is clear: this world is God's, entrusted to us. He will hold us accountable for our stewardship. We should be in the vanguard of the environmental movement. Instead, most Christians treat the world as merely the stage for God's saving work and our daily life. We regard Christian environmentalists with amusement or even contempt as misguided enthusiasts diverted from their main task of preaching the gospel with love and urgency. How tragic this is! How unscriptural!

Environmental concerns have surfaced time after time in history. One of the main causes of the fall of the Babylonian Empire was the collapse of its agricultural system through over-irrigation. The Mediterranean area was once a major cereal growing area, but is now severely degraded. China, Japan and much of temperate Europe are largely deforested. Famines have recurred—and still occur—throughout the centuries. Time after time we have contributed to our own problems. The early Polynesian population of New Zealand cleared so much forest that the large flightless moas[10] that were a key item in their diet were virtually extinct within a few centuries. Land misuse and conflict almost led to the extinction of the Easter Islanders. The Dust Bowl of south-central U.S. arose from over-intensive cultivation in an area where rainfall is low and the soil fragile. The list is long.

Those affected by problems in their environment naturally take action. At first, problems are local: drought, pollution, over-grazing. For centuries, the solution was to move on, as evidenced in slash and burn agriculture, Viking colonization and European expansion into the Americas and then Australasia. But we are now running out of world. There are ever fewer places to find sanctuary. Human movements have turned into refugee problems. Economic opportunism is increasingly giving way to environmental desperation.

For our part, we have neglected our calling. "Environmentalism" is now driven by secular concerns. As citizens, we may see these concerns as important, expressing our worries about deforestation, ozone depletion, changing climatic patterns, and the rest. But we do not treat them as Christian issues. They score poorly on our list of spiritual priorities. I hope it is now clear that such an attitude is non-biblical; that God has laid on us environmental responsibilities as well as privileges, and to shirk these responsibilities is morally wrong or, in Bible language, sinful.

However, it is unhelpful merely to look back. I believe there is, or should be, considerable convergence between Christian and secular ways as far as our environment is concerned. We should ponder seriously whether we ought not to make common cause with our environmentalist brethren, and perhaps even take the lead in helping them in the moral challenges that the world has to face.

There is a choice before us. It is an old, old choice: God's way or our way. God's way is creation care, responding to both his Book of Works and to his Book of Words. Do we have enough courage and confidence to say "yes" to him in every part of our life? Or are we going to be selective in our responses, saying "yes" in personal and social matters but closing our eyes to the world around us—God's world?

On Christmas Eve, 1968, the Apollo astronauts in orbit around the moon read to the listening world below them the first ten verses of the Book of Genesis. Frank Borman then prayed:

> Give us, O God, the vision which can see thy love in the world in spite of human failure. Give us the faith to trust thy goodness in spite of our ignorance and weakness.
>
> Give us the knowledge that we may continue to pray with understanding hearts.
>
> And show us what each one of us can do to set forward the coming of the day of universal peace.[11]

As we face the future on this battered planet, we can do much worse than to make this our own prayer.

---

[1] World Commission on Environment and Development chaired by Gro Harlem Brundtland, *Our Common Future* (NY: Oxford University Press, 1987), p xi.

[2] As far as humans are concerned, this instruction may be regarded as having been terminated by the New Covenant. God's original care and nurture were concentrated on his chosen people from Abraham on. However, the Old Testament genetic line has been superseded by a spiritual line. The command to "fill the earth" with our own descendants has become an evangelistic charge for spiritual increase through "making disciples of all nations."

[3] Ian L. McHarg, *Design with Nature* (New York: Wiley, 1969), 197.

[4] Jacque Delors, "Opening Address," in *Environmental Ethics,* eds. Philippe Bordeaux, P.M. Fasella, and A. Teller, (Luxembourg: Office for Official Publications of the European Communities, 1990), 22.

[5] C.E.B. Cranfield, "Some Observations on Romans 8:19-21," in *Reconciliation and Hope: New Testament Essays on Atonement and Eschatology Presented to L.L. Morris on His 60th Birthday* (Grand Rapids, MI: Eerdmans, 1974), 227.

[6] D. Kidner, *Genesis* (London: Tyndale, 1967), 73.

[7] This verse tells of judgment: "Your [God's] wrath has come, and the time for . . . destroying those who destroy the earth" (NRSV).

[8] Henri Blocher, *In the Beginning* (Leicester, UK: Inter-Varsity Press, 1984), 184.

[9] An excellent debunking of this idea is given by University of California Professor Daniel Botkin in *Discordant Harmonies* (New York: Oxford University Press, 1990).

[10] Especially *Dinornis giganteus* , which stood over two meters high at the shoulder.

[11] Borman's prayer may be found on the Web at http://www.friends-partners.ru/partners/mwade/flights/apollo8.htm

# 3

# Ecology and Christian Ethics

*Michael S. Northcott*

## Opening scenario

In a remote forest in the hills on the border between the State of Sarawak, East Malaysia and Kalimantan Indonesia, in what used to be called North Borneo, a Malaysian Chinese logging company extracts lumber from a forest that has been standing there for many tens of thousands of years. The company has a concession from the state government, for which they have paid handsomely, and their job is to cut down the largest trees, drag them to the river and float them to the saw mills, plywood plants and docks in Sibu. Some logs will be placed whole onto barges and shipped further east to Japan where they might end up as disposable chopsticks or as disposable casings for concrete moldings in construction work.

The process of timber extraction is instructive. The machinery is shipped in by river and a path is cleared in the thick forest for men and machines to reach the largest trees, which is their objective. Two men operate a huge mechanical saw and cut through the hard and heavy wood of a majestic tree that is typically 200 feet high and 25 feet around. As the tree falls, it takes with it smaller and thinner trees and a mass of epiphytes (creepers, ferns, orchids and other plants that have been growing on it). It also takes with it thousands of insects that have inhabited it from its root system—including large above-ground buttresses spread over the thin soil of the forest floor— to its leafy canopy more than 150 feet in the air, where birds will have fed on the abundant insect life.

Once the tree hits the ground, men remove the branches and creepers from the trunk, attach ropes and cables around its great length, and drag it down to the river. To create a path for the trees the loggers cut a large number of smaller trees and lay them side-by-side to act as a road. This timber road is then lubricated with diesel oil to make it easier to drag the tree to the river. In the water, it is lashed into a raft with other logs to be floated off to a mill or dockside.

After only a small number of trees have been extracted by this method, a mature stand of ancient forest is reduced to an appalling state from which it may never fully recover. This is because tropical forests are highly complex ecosystems, with the earth's greatest variety of plants and animals per square mile. Indiscriminate logging not only decimates tree populations, but it can profoundly disturb a whole host of species that live in a symbiotic relationship with the extracted trees and each other.

In the building where I am writing this paper, a new carpet was laid a few years ago by outside contractors. To provide a smooth base and so extend the life of the carpet, fitters installed a new sub floor of plywood. As they delivered the timber, I noticed that the wood had been made in Indonesia. Under the floor of this fine Victorian building lies some of the timber extracted from the forests of Borneo, steamed into a wood-and-glue sandwich—all to extend the life of a polypropylene carpet.

The ecological crisis is a global as well as a local phenomenon in which we are all caught up. Our hands may not be on the chain saws. Our eyes may not watch the dials that monitor the toxic outputs of large chemical plants. We may not be directly responsible for decisions to locate toxic waste dumps and incinerators in poor neighborhoods. Yet our homes, offices and cars are made as products of modern industry. For more than two hundred years, manufacturing has transformed the astonishing and myriad diversity of the living species of the planet—together with the oils and minerals their forebears laid down under the surface of the earth—into organically dead consumer artifacts.

As a consequence of the scale and reach of industrialism, tribal, cultural and language groups, as well as non-human species, are disappearing at rates unprecedented in planetary history. Indigenous tribal groups suffer tragically from the destruction of the forests and rivers in which they have foraged for tens of thousands of years. They may end up in rural or urban slums, not comprehending the forces that shifted them. Many others succumb to the diseases that loggers, miners and farmers introduce to their remote regions, against which they have no resistance. Displaced mammals die in large numbers when they are hunted to extinction, burned alive in forest fires or starved when their foraging land disappears.

## Is religion to blame?

The Jewish and Christian religions are often blamed for the global ecological crisis because of their shared inheritance of the concept of "dominion" as introduced in the account of the Garden of Eden and because of a doctrine of creation which represents God and creation as separate spheres of being. However, the causation of the ecological crisis is in reality deep and multi-factorial and cannot be reduced to the influence of one doctrine or concept or religious tradition. Environmental historians point to the rise of modern science and technology, to the growth of industrial methods of production, and to a global market in natural resources and manufactures as key causative elements in the crisis. Other roots of the crisis that historians and environmentalists discern include

- ↪ patterns of gender oppression and domination that impacted humanity's relation with other species in classical as well as Christian thought;

- ↪ Renaissance affirmations of "man" as the measure of all things and a consequent loss of respect for God as Creator and for the created order as providentially ordered by God;

- ↪ the rise of Western individualism and materialism with its growing focus on material achievement in this life as the locus of all human goods and purposes, rather than as preparation for the life to come.[1]

Ecological and environmental responses to the crisis have been as varied as its root causes. One of the principal strands of ecological thinking, deep ecology, involves an attempt to recover a more holistic concept of life on earth, in which human life and society are reconnected with the biophysical character and limits of ecological systems. The moral imperative in this view is for human activities to be designed, but also restrained, so as to preserve the health and diversity of these systems.[2] In economic thought, for example, this approach involves the proposition that the human economy is a sub-system of the physical economy of the eco-sphere. Physical limits to economic growth and to the quantity of money in circulation are seen to arise from the limited quantity of natural resources on the planet, and the limited capacities of the atmosphere, oceans and land areas to absorb the waste products of industrial production and economic growth.[3] This approach presupposes that the earth and its living systems are sources of intrinsic value and that actions to conserve the life communities of the planet are therefore a prime ethical impera- tive for humans.[4] Ecological holism in its deepest forms involves a project to re-sacralize[5] the life of the earth, to discern spiritual forces within its living systems or even to conceive of the earth as a goddess, sometimes called *gaia*, after the Greek earth goddess.

An alternative approach involves the reform, rather than the over- throw, of the dominant utilitarian and human-centered frame of con- temporary ethical discourses. Advocates of this approach argue that although moral value is located in human propensities and goods, such as the human capacity for happiness, these goods should be pursued in such a way as to minimize harmful consequences for other species and ecosystems. This is so that future generations may con- tinue to enjoy the fruits of the earth as we, or at least as the inhabit- ants of the more affluent countries of the world, currently do.[6] De- vices for achieving such ends include new forms of environmental accounting whereby the full ecological costs of the production of par- ticular goods and services are accounted for in price regulation and taxation systems.

## Christian environmental theology

Environmental theologians also adopt a range of approaches to the ecological crisis. Many theologians acknowledge that although Christianity may not be to blame for the crisis, features in the history of Christian theology, and in particular doctrines of creation and re- demption, may play an important part. Biblical writers and early Christian theologians such as Irenaeus placed a particular emphasis on the integral relationship between God's purposes for creation and for humanity, and regard the redemption wrought by God in Christ as of significance for the whole created order, and not just for hu- mans. Theologians since Augustine, however, have come to place a much greater emphasis on human uniqueness, and to focus the doc- trine of redemption on the human condition, and especially on the salvation of the human soul. This more disembodied and anthropo- centric concept of salvation has led to salvation schema that exclude animals and the non-human creation from the eschatological vision of redemption and focus narrowly on the spiritual, and rational, na- ture of humans as the principal object of divine concern within cre- ation.

The recognition of the anthropocentric turn in the Western Christian tradition, particularly as it began to absorb key elements of Greek and Roman philosophy and culture from the fourth century, has led some theologians to a wholesale rejection of traditional Christian metaphysics and creation theologies. Many North American eco-theologians have adopted the metaphysics of Alfred North Whitehead's "process" philosophy. They argue that evolutionary and ecological insights require us to re-conceive nature, humanity and God as enmeshed in a set of creative life processes whose outcome is indeterminate or not predictable. The goals of life emerge within the process of life itself and the purposes of God for humans and the biosphere are caught up in these uncertain outcomes.

This theological position involves a kind of pantheist, or panentheist, stance in which God is said to be hidden within the "processes" of the history of being.[7] According to its advocates, this approach invites Christians to re-sacralize the world and all of its species as the "body of God" and hence, it is argued, to treat the natural order with new respect.[8] As with the holistic stance of deep ecologists, process metaphysics is said to involve a new radical ethical valuation of the intrinsic worth of all living systems and their individual inhabitants. It is connected by some writers with a radical theological biocentrism in which every living thing is said to have as much value to the God who is hidden in the processes of life as every other.[9] Instead of human dominion, this approach commends a planetary egalitarianism in which human comfort, even human survival, must be set in an ethical balance with the needs, and goals, of all other living things.

Critics of deep ecology, and of process-oriented eco-theology, point out that the ethical outcomes of these approaches are by no means unambiguous. Positions that merge human and non-human identities (and for some divinity as well) require the adoption of monistic ontology, making it hard to distinguish the respective goods of different kinds of being and groups of species. This kind of merging of identities removes the ground for recognition of the Other as that which is not Self, and hence as deserving of recognition and respect independently of the desires and interests of the Self. In a monist perspective, a particular act may have ecologically destructive consequences. But provided the act may be said to enhance the life quality of those who engage in it, it may be said to be ethically justified since these persons may be said to be "one" with that which they are destroying. They will thus further the fulfillment of the totality of being by their ac-tions.[10] Monistic approaches provide weak grounds for discerning be-tween good and evil actions, and for resisting evil. Evil outcomes are presented by process theologians as simply the other path that life takes at different points in the process. The idea of the ordering of creation towards the good purposes of a good God, or the possibility and legitimacy of resistance to evil through participation in such a God, is alien to this perspective.

More orthodox environmental theologians have sought to revisit the Jewish and Christian traditions. They've suggested that within these traditions lie important resources for an ecological worldview and a Christian environmental ethic, which, though neglected in re-cent centuries, can be restored.

The central idea in this restoration is the doctrine of creation, and in particular the concept of the biophysical world as the creation of a good God. According to the Creation narratives in Genesis, humanity is bequeathed the earth not as a possession but as a gift. The word "dominion" implies not ownership but trusteeship or stewardship. Just as the Israelites were not owners of the land of Israel, but received it as a gift from God, so Adam and Eve, the archetypal ancestors of Is-rael, are conceived as tenders of the garden of the world, not its abso-lute possessors. Therefore, when they ignore the maker's instructions, they are removed from the Garden. They were there on trust. The clear implication is that humans are given the earth on trust, to steward it on behalf of its Creator and Lord. The Israelite theology of land re-flects this view. The land of Israel, "flowing with milk and honey," was given by the Lord to Israel as a part of their redemption from slavery. The Israelites were to tend it so as to preserve its fertility. They were not to use their access to land in such a way as to oppress the land-less, or exclude wild animals from it, nor were they to till the soil over-harshly.

## Covenant and the environment

The concept of land as gift, and of its original and continuing di-vine ownership, is connected with the covenant between Yahweh and the people of God. Robert Murray points out that the word "covenant" is used by Jeremiah to speak of God's original ordering of the space and time of Creation and of God's blessing of the line of David (that ruled Israel), and the tribe of Levites, (who ordered the worship of

Israel) (Jer. 33:20–26[11]). The covenant between God and the people of God is in effect a cosmic covenant that has significance both for the ordering of relations between God and God's people, for their own society and for their relationship with the rest of creation. The covenant community embraces humans, non-humans and the land itself. And so the first duty enshrined in the covenant—to worship the Lord—is a duty that is shared by both humans and non-humans, and even the land and the sea and the elements. As the Psalmist says: "Praise the Lord from the earth, you sea monsters and all deeps, fire and hail, snow and frost, stormy wind fulfilling his command!" (Ps.148:7–8, NRSV).

This cosmic concept of covenant involves the idea of a deep relationality that is set into created order between the cosmos and human vocations of worship, food production and other forms of social organization and exchange. This is reflected particularly in the kingly Psalms, such as Psalm 72 (vv. 1–6) where an explicit relationship is described between the justice and righteousness of the society over which the king presides, the ordering of day and night in the cosmos, and the fertility of the land. Where God is acknowledged as Lord and where the people participate in worship and model their lives after God's righteousness, the land will prosper and its fruitfulness will be maintained. Where people neglect the terms of the covenant, worship that which is not the just and holy God, and allow some to grow wealthy and acquire great parcels of land while others languish in poverty, the land itself will turn to desert and its non-human inhabitants will also suffer. As Isaiah puts it:

> The world languishes and withers; the heavens languish together with the earth. The earth lies polluted under its inhabitants; for they have transgressed laws, violated the statutes, broken the everlasting covenant. Therefore a curse devours the earth, and its inhabitants suffer for their guilt. (Isa. 24:4–6, NRSV)

This is a powerful ecological insight, for it connects human injustice with ecological destruction, and there is much evidence in the history of the environmental crisis that accords with this connection. After the Clearances in Scotland, much of the land that had formerly been forest, market garden and land for animal husbandry was turned over to sheep and deer herds. This was in order to support the incomes and sports of the Lairds who began to aspire to the wealthier

lifestyle of their English baronial partners after the Act of Union. Today, much of the land of the Highlands is degraded. It supports very
few species and suffers from serious water and wind erosion. Over-
grazing by sheep and deer reduced the natural forest cover and the
land could no longer, in its current condition, support the many thousands of tiny rural communities that once lived off it. Similar ecologically destructive practices have cleared tribal peoples from tropical
rainforests where, before being supplanted by loggers and cattle
ranchers, they had derived a good and dignified life for many thousands of years. The neglect of natural justice by governments and corporations that trample on the land claims of indigenous peoples leads
both to human and ecological disaster.

The idea that there is a deep relationality of God, humans and the
earth within created order is given a much deeper significance when
read in the light of the events of Christ's life, death and resurrection.
In the Gospel of Luke, the first words of Christ's public ministry an-
nounce "the year of the Lord's favor" (Luke 4:19). They refer to the
Jubilee year, when lands were given back to families that had lost them
through debt, and the original equitable distribution of the land to
the families of the tribes of Israel was restored. In the nature miracles,
Christ is recognized by earth and sky as the Lord of the cosmos. The
wild animals are said to have ministered to him in his 40-day sojourn
in the wilderness. During the crucifixion of Christ the sky is said to
have become dark and the earth to have quaked. The salvation and
forgiveness of sins that the Crucifixion and Resurrection brought near
are also described in cosmic terms in the Gospel of John, in the Pauline
Epistles and by the theologians of the early church.

## Resurrection, ecology and the cosmos

The resurrection of Jesus Christ is central to an orthodox approach
to ecological ethics. It is the resurrection of the material and living
body of Jesus that the church celebrates, not, as the Greeks do, the
resurrection of the immaterial soul.[12] In the Resurrection, God turns
back the effects of sin on the material, embodied, order as well as the
rational and spiritual modes of life. The relational alienation that characterizes the state of sin—between God and humans, between persons, and between humans and the non-human creation—is overcome. The embodied created order is reoriented toward its original
destiny of participation in the divine:

For in him all the fullness of God was pleased to dwell,
and through him God was pleased to reconcile to himself all things, whether on earth or in heaven. (Col. 1:19–
20, NRSV)

From the perspective of the resurrection of Christ we can see again
the original destiny and goal of the whole created order:

Then I heard every creature in heaven and on earth and
under the earth and in the sea, and all that is in them,
singing, "To the one seated on the throne and to the
Lamb be blessing and honor and glory and might forever and ever!" (Rev. 5:13, NRSV)

The implications of this cosmic reading of the life, death and resurrection of Jesus are profound, and were taken up with alacrity by
the second century theologian Irenaeus of Lyons. Against the Gnostics,
who regarded matter as independent of God and a counter-force to
divine life in the cosmos, Irenaeus developed a deeply materialist and
relational account of creation. He argued that the fullness—the
*pleroma*—of God included all things and that nothing exists that is
unrelated to God, who contains all things and commands all things
into existence "by his Word that never wearies."[13] Irenaeus saw the
relationality of all things within the cosmos as modeled on the original relationality of the triune Creator. The fall of humanity marred
this original relationality, but in the birth of the original creative Word
into human flesh and matter, the goodness of the material world is
restored through the coming of Jesus Christ.[14] Drawing on the Revelation of St John the Divine, Irenaeus conceives of this restoration in
terms of a final recapitulation of the whole embodied creation at the
end of history in which the earth is physically transformed in the final echelon.

This cosmic reading of the salvation and resurrection of Christ is
increasingly abandoned by later theologians, however. In the writings of Athanasius, Augustine, Aquinas and Calvin we find a much
more anthropocentric concept of salvation. The influence of the Greek
philosophical tradition refocuses the doctrine of redemption on the
souls of rational creatures—humans—while other animals, and the
created order, are increasingly viewed as means to the ends of the
higher order of humans. Later theologians, particularly in the Western tradition, placed much greater stress on the idea of human dominion over creation, and so prefigure the Renaissance and modern

dogma of the supremacy of man. In this sense, Lynn White's well-known ecological complaint against the Christian tradition has some limited validity.[15] Yet it is not the original tradition that is at fault, but rather the classical overlay of the tradition that pushed it in a more rationalistic, idealistic and hence disembodied direction. For John Calvin, this anthropocentric approach meant that only in the souls of the elect was it possible to discern an unambiguous reflection of the divine light, which at Creation irradiated the whole universe.[16] And according to Calvin, the ethical vocation of the elect is to work for the material transformation of creation, putting it at the service of humans, so that, by adding to their glory, it may at last be redeemed.

It is not hard to see why Max Weber should have regarded this ethic as being closely related to the spirit of capitalism, with its focus on the vocation of individuals in their secular work to reorder the creation in the pursuit of human goods. Nor is it difficult to see a close connection between this kind of Christian anthropocentrism and the rise of modern ethical individualism. In ancient Jewish and Christian metaphysics, the whole created order is conceived as a moral order. Its form and beauty reveal a moral patterning that, after the first and second chapters of Paul's Epistle to the Romans, Christians came to describe in terms of a "natural law." Following the commandments of God, which are said to be the revealed form of the natural law, involves an intrinsic respect for created order that issues in particular from the first commandment to worship God and not to put any created thing in the place of God. In anthropocentric revisions of creation theology, this cosmically located moral order is subdued in favor of a focus on the moral status of the souls of the elect.

This moral focus on the individual is taken up with particular force by Enlightenment philosophers who judged that the locus of moral value is the mind of rational creatures, and no bodies or things that are not rational may be said to have moral significance or to express moral agency. Natural law gives way to the two dominant strands of modern ethical discourse: ethical intuitionism, also known as hedonism or utilitarianism; and deontological ethics. Both strands focus on the human person as the sole locus of moral value, and on life in this age to the exclusion of the next. The difference between these positions is in the mode of derivation of value from persons. In the case of utilitarianism there is an emphasis on inner moral intuitions and emotive states. Those actions that have the consequence of enhancing human emotive states are said to be ethically right. In a

deontological perspective it is the person *per se* that is the ethical focus, rather than the emotional states of the person. Actions and intentions are judged not according to outcomes but according to their
treatment of persons. Actions that treat persons as ends are inherently good; actions that treat persons as means are not.

## Mutuality and interdependence

Both of these ethical traditions are problematic in ecological perspective for they both place undue emphasis on persons at the expense of moral values located in the larger created order. Christian
tradition locates moral value not in persons or in features of created
being in the first instance, but in the being of God. God as original
loving community of being generates the Creation out of the generosity of divine love. God loves the creation and sets a relational order
of mutuality and interdependence deep within the created order that
reflects God's own relational being. We are given glimpses of this relational order in the being of God in the Hebrew Bible. But it is not until
the incarnation of Jesus Christ that the full relational character of God
as Trinity, and hence the Trinitarian and relational conception of the
community of God and creation, is revealed. Jesus Christ as the first
principle, or *logos,* of creation appears within the material and relational web of creation: "He came to what was his own, and his own
people did not accept him" (John 1:11, NRSV).

Moral value in this view is shared by the whole creation, and not
just by rational or ensouled beings. God loves matter as well as spirit,
and God's intention in creation, restored in redemption, is to draw
matter as well as spirit into the original love and generosity of divine
being. Humans have abused their freedom and marred this original
loving ordering of the creation. God in Christ restores this original
ordering and makes possible the kind of redeemed relationships between humans and the rest of creation that we see in Isaiah's vision of
the peaceable kingdom where "they will not hurt or destroy on all my
holy mountain" (Isa.11:9, NRSV). This is also seen in the Gospel picture of Christ as the cosmic Lord being ministered to by the animals
and holy angels.

The Christian spiritual tradition is replete with stories of wise men
and women who had a particular affinity for wild spaces and wild
animals, from St Anthony, the founder of eremitic monasticism, to St
Jerome, St Isaac the Syrian, St Hildegaard of Bingen and St Francis of

Assisi. In our own century, some of the most prominent environmentalists have also been people with a deep rooting in the Christian spiritual tradition. Early in the century, and long before the animal rights or environmental movements commenced, Albert Schweitzer preached some powerful sermons on reverence for life, and in particular on compassion for animals, which were later taken up in Karl Barth's exposition of the covenantal ethics of life in the *Church Dogmatics.*[17] John Muir wrote dazzling accounts of his wilderness wanderings and of the divine spirit he found imbued in the Sierra Nevada, which he subsequently devoted his energies to preserving from industrial predation.[18]

## Christian environmental ethics

In the light of this ecological re-reading of the Christian tradition, and in light of the now dominant tradition of secular ethics, it is possible to outline a Christian approach to environmental ethics as follows:

**1. A Christian environmental ethic involves as its first premise a confession of human sin and guilt for the desecration of God's earth, and recognition that without the redemptive incarnation, crucifixion and resurrection of Jesus Christ, neither creation as a whole, nor humans within creation, are capable of recovering true relationality with the being of God.** Similarly, the restoration of a just and respectful relationality between humans and created order depends upon the spiritual and embodied events of salvation that are focused in the life, death and resurrection of Christ. Confession of ecological sin, and the recognition of Jesus Christ as the locus of ecological salvation, also involves a recognition of the anthropocentric orientation that Christian theology, and, under its influence, secular modernity, have taken in the past. As earlier Christians recognized, and often expressed in stone, carols, icons and theology, on the holy tree of the cross of Christ the creation is healed and restored, and humanity's sin, including sins against creation, is forgiven. Reconciliation and forgiveness are fundamental features of Christian ethics. Without reconciliation, we continue to inhabit an order characterized by sinfulness. It is also an essential element in a Christian environmental ethic. Environmental apocalyptic is very good at pointing the finger of blame and generating guilt. And judgment and guilt can so easily lead to denial, an inability to own shared human responsibility for environmental destruction and so to a refusal to change.

**2. As the first premise of a Christian ecological ethic involves worship—confession and acknowledgement of the cosmic Christ as the locus of humanity and creation's healing and reconciliation—so the second also involves worship.** The praise of God as creator, governor, upholder and sustainer of all living things is the first moral duty and vocation of all life according to the summary of the law taught by Christ in the New Testament: The First Commandment is this—love God with all your heart, soul, mind and strength (Mark 12:29-30). Acts of praise and worship that draw upon the Hebrew Psalter will include frequent reference to the community of creation gathered in praise at the altar of God. The Psalmists see the human response to God in praise as part of the larger response of the whole community of creation to the generosity of God:

> Let the heavens be glad, and let the earth rejoice; let the sea roar, and all that fills it; let the field exult and everything in it. Then shall all the trees of the forest sing for joy before the Lord. (Ps. 96:11–13a, NRSV)

Worship is the first response of humans and of the whole creation, and it is the spring from which all other ethical duties arise for Jews and for Christians.[19]

The Sabbath, which is set aside for worship, has particular ecological significance. It sets limits on human work in creation; it commands a rest for all creatures and not just for humans. According to later Jewish interpreters of the Sabbath law, money is also supposed to rest on the Sabbath, and not to gather interest. In other words, worship of the Lord sets limits on the capitalist remaking of the world after human desires and comforts. Worship of the Lord is not just a Sunday affair, however. The monastic tradition of the daily Divine Offices has sustained an elitism in Christian spirituality that requires religious and secular clergy to say the office daily (an office that draws deeply from the Jewish Psalter) while lay people are encouraged—but not obligated—to devote a few minutes each day to devotion. If worship it to genuinely involve a spiritual re-ordering of desire towards those goods for which persons are made—relational participation in God, with other persons and with the rest of creation—Christians need to daily encounter a contemplative tradition where work in the world, and enjoyment of its fruits, is set in the context of contemplation of the beauty and order of the divine nature that is reflected in the beauty of creation.

Re-engagement with the natural world, particularly for urban people, may also play a part in the modern recovery of a more contemplative and conserving approach to creation. Deep ecologists such as Arne Naess identify peak experiences in the wild as means to the healing of the distorted modern psyche from excessive attachment to material success, ambition and ownership of consumer goods.[20] But there is a danger of nature fetishism in this kind of approach. It can also signal a denial of the larger systemic character of the sinful actions and systems that are degrading the natural order, as one half of humanity is caught up in a mad dash for unprecedented wealth while the other half is oppressed by environmental poverty and degradation. There is also a danger of substituting the worship of the creature for worship of the Creator. Paradoxically, an orientation towards God in meditative prayer is ecologically fruitful partly because it qualifies the quest for fulfillment in material and mental satisfactions in this life. If people come to own that their highest good is union with God then they are much less likely to spend their energies and resources pursuing ecologically damaging forms of material fulfillment in this life.

**3. Worship is not a solitary act.** Christians worship as members of local communities of worshipers and their acts of worship involve relational affirmations of our mutual dependence one on another. As the writer of the First Epistle of John says, "Those who love God must love their brothers and sisters also" (1 John 4:21). Equally, no spiritual or contemplative experience of God or of wilderness can be said to be ethically fruitful if it does not issue in more virtuous attitudes and behaviors, towards God, other persons, and the creation. The biologist Edward Wilson advances the concept of biophilia in his recent writings, but he sees love of nature as a purely human emotion and relation between persons and nature.[21] James Nash and Sally McFague suggest that a central feature of a Christian ecological worldview is love for nature. But instead of being solely informed by fragmented human experiences of love, it is shaped by the revelation of the love of God that is definitively revealed to humans in Jesus Christ.[22]

Love in the Christian tradition is also said to be a fruit of participation in the Spirit that characterizes sharing in the divine nature into which Christians are invited by Jesus Christ. Participation in God is the foundation for the renewal of relations between Christians in the body of Christ. This renewal is inaugurated by the events of Christ's life, death and resurrection and is brought near through participation in the life of the Spirit, and in the body of Christ as the spiritual community wherein this life finds expression on earth and in the present age. In this spiritual community a new order of relationality begins—a new fellowship is established, in which the walls of division that once alienated persons from one another are broken down. In the Christian spiritual tradition, this new order is seen at many points to have significance beyond human relationships for the relations of the saints with the whole creation.[23]

**4. Relationality finds classical Christian ethical expression in the practice of the virtues, and it is in virtue ethics that the ethical significance of worship and community is most clearly expressed. Love, fidelity, justice, peaceableness, temperance, prudence, courage and hope are all central to the recovery of an ecological ethical framework for human living.[24]**

Love of God and love and respect for life are deeply interconnected. It is hard to love something that is the product of a random chain of events. Love for the giver and love for the gift are intertwined. And love requires recognition that the Other, including the other beings of creation, is not Self, and has different needs from the Self that require respect. In other words, love for nature is not so much a warm inner feeling as it is the recognition of the needs of our non-human neighbours with which we share space but who have different needs and goods. Human love for nature is more likely to have ethical significance when it is expressed in relation to those particular places in the natural world with which individuals and communities have some connection, or which are geographically close to where they live or work.

If all local communities were active and empowered in preserving those places in the natural world that they live close to, then every place would have its lover, its conserver, its defender against the predations of developers, loggers, road builders and the toxic outflows of our industrial society. Communities of place that are also worshiping and ritualizing communities have a good record of environmental conservation, from the Hindu tree huggers of Rajasthan to the Free Church defenders of the Isle of Harris who fought against a proposed super quarry.[25] Ecological politics is often presented as a global phenomenon involving inter-governmental treaties, conferences and regulations. But it is where groups of people and communities have been faithful to their own local environments, loving, stewarding, and defending them, if necessary, in the courts, that environmentalism has been its most successful. Here is a prime role for churches in the environmental and conservation movement. Churches remain the largest single form of voluntary organization, even in secularized societies such as those of northern Europe. They are an enormous potential resource for environmental fidelity. Worship and ritual that rebuild the connections between love for Creator and love for creation will have a significant impact on the environment our children and their descendants will inhabit.

**5. Justice is also a core environmental virtue.** *Sedeq*, divine righteousness, is understood in the Hebrew Bible to be particularly focused on the victims of injustice: widows, orphans, the poor. There are more refugees from environmental disaster than from any other cause in the world today, including war. It is also the case that in Europe, North America and elsewhere, the most polluting factories, incinerators and toxic waste dumps are frequently located in the neighborhoods of poor people. Similarly, many particularly polluting or dangerous industries have been exported wholesale from wealthy nations in the northern hemisphere to poorer nations in the South. The Hebrew Bible makes a clear connection between injustice in human society—and in particular, excess wealth accumulation by the greedy few—and ecological destruction. This same connection is evident on a global scale on the cusp of the third millennium. The world has never had so many extremely wealthy people. Nor has it ever seen so many hungry, so many living in inadequate housing, so many forced to degrade their environments in order to stave off starvation. Recognition of the connection between environmental crisis, social injustice, and human sinfulness is a central theological and ethical insight. The cultivation of a passion for justice is correlatively an essential feature of an ecological ethic.

The first place where Christians may seek ecological and social justice is in their own organizations. Churches that are invested indiscriminately in global stock markets, and that are large landowners and use their land to maximize rents rather than to care for the earth are in a poor position to preach ecological justice to the secular world. Systemic economic reform is key to the reorientation of industrial economies towards the physical limits of planetary ecosystems. Herman Daly argues that mechanisms to limit wealth inequality, in particular, are at the heart of ecological economics and constitute what he calls an "eleventh commandment."[26]

The quest for justice is also the quest for original peace—the peaceable kingdom. Recognizing this allows Christians to continue in the struggle for ecological justice without being downhearted when it does not arrive the following week. As well as carrying this eschatological orientation, the virtue of peaceableness also has important social, political and ecological implications. The Hebrew historians and prophets condemned the more worldly of the kings of Israel for idolizing other gods, and for adopting the values and styles of power and domination that these gods stood for instead of the justice and compassion of Yahweh. The armies of ancient Mesopotamia, including at times the armies of Israel, stripped the land bare of its fertile soil and much rich agricultural land was over-intensively cultivated and hence desertified.[27] As well as persons, creation itself is a major casualty of war and of the diversion of resources for war preparations.

**6. Temperance and prudence are also core environmental virtues.** Temperance involves us in moderating our demands on nature. Stephen Clark argues that the real cause of the environmental crisis is the excessive modern quest for comfort and satiety.[28] Intemperate desires are regarded in many spiritual traditions as the consequence of a spiritual vacuum, and a recovery of spirituality may therefore be said to be central to the ecological conundrum.

Prudence is of crucial importance in a global civilization where new technologies, such as genetically modified foods or mobile telephones, can spread from a small experimental base into a mass practice in a very few years. The precautionary principle is a key ethical device for restraining the science-led urge to advance risky technologies into the mass market before their full ecological consequences have been tested and explored.

**7. Courage is also an essential ecological virtue.** Courage is called for in resisting the spirit of the age that sets material acquisition above the fair distribution and preservation of the fruitfulness of the earth. Courage is also required in challenging powerful corporations and institutions that anonymously subvert and undermine the sustainability of ecosystems near and far. In the New Testament, Christ commends non-violent resistance to evil, a kind of resistance that shames the evildoer and reveals evil for what it is.[29] It also requires courage not to close the mind to the enormity of the ecological crisis, and to one's daily connections with it, and to sustain daily acts of ecological resistance against social pressures. In such acts of resistance, environmental ethics become real: eating less meat, buying fewer clothes and electronic items, walking or cycling instead of using a vehicle whenever possible, cutting down on foreign travel, turning down the heating.

## Closing word

An environmental ethic is not all about courageous resistance and asceticism. It is also about rejoicing in alternative forms of comfort and joy that have low environmental impacts but high satisfaction scores; in particular, people gatherings such as worship, eating, socializing, music-making, story telling, artistic performance, nature walking and joining together for political and ecological engagement. Communities that live from the relational strengths of such actions will be communities of hope, especially when such actions have as their focus the hope for the restoration of all created life that Christians find in the resurrection of Jesus Christ. This is the source of hope for Christians; an ecological ethic does not involve a quest for an impossible utopia. It can be sustained in worshiping communities that practice ethical resistance to the dominant morés of consumer society.

---

[1] See further my account of the causation of the environmental crisis in Michael S. Northcott, *The Environment and Christian Ethics* (Cambridge, UK: Cambridge University Press, 1996), ch. 2.

[2] Aldo Leopold's *Sand Country Almanac*, paperback edition (New York: Oxford University Press, 1968) has become a classic in the elaboration of this kind of holistic ecosystem ethic, also known as the land ethic.

[3] See further Herman Daly, *Beyond Growth: The Economics of Sustainable Development* (Boston: Beacon Press, 1997).

[4] For a full exposition of the concept of intrinsic value, see further Holmes Rolston, *Environmental Ethics: Duties to and Values in the Natural Environment* (Philadelphia: Temple University Press, 1988).

[5] To make life holy or sacred again.

[6] For a classic exposition of this reformist position, see Robin Attfield, *The Ethics of Environmental Concern*, Second Edition (Athens, GA: University of Georgia Press, 1991).

[7] See further Jay B. McDaniel, *Of God and Pelicans: A Theology of Reverence for Life* (Louisville, KY: Westminster/John Knox Press, 1989) and Charles Birch and John B. Cobb, *Liberating Life: From the Cell to the Community* (Cambridge, UK: Cambridge University Press, 1981).

8  See especially Sally McFague, *The Body of God: An Ecological Theology* (London: SCM Press, 1993) and Grace Jantzen, *God's World, God's Body* (London: Darton, Longman and Todd, 1984).

9  Matthew Fox appears to commend this kind of biocentric egalitarianism in his *Original Blessing: A Primer in Creation Spirituality* (Santa Fe, NM: Bear and Company, 1983).

10 See further Val Plumwood, "Nature, Self and Gender: Feminism, Environmental Philosophy and the Critique of Rationalism" in *Environmental Ethics,* ed. Robert Elliot, (Oxford, UK: Oxford University Press, 1995).

11 Robert Murray, *The Cosmic Covenant: Biblical Themes of Justice, Peace and the Integrity of Creation* (London: Sheed and Ward, 1992), 4.

12 For a full exposition of the centrality of the Resurrection to Christian ethics, see Oliver O'Donovan, *Resurrection and Moral Order: An Outline for Evangelical Ethics* (Leicester, UK: Inter-Varsity Press, 1986).

13 Irenaeus, *Against Heresies,* II, 2.5.

14 See further Northcott, loc. cit., 208–9. See also Colin Gunton, *The One, the Three and the Many: God, Creation and the Culture of Modernity* (Cambridge, UK: Cambridge University Press, 1993).

15 Lynn White, "The Historical Roots of Our Ecologic Crisis," *Science,* 155, 10 March, 1967, 1203–7.

16 John Calvin, *Institutes* II, 2.19.

17 Albert Schweitzer, *A Place for Revelation: Sermons on Reverence for Life,* translated by David L. Holland (New York: Macmillan, 1988).

18 See John Muir, *My First Summer in the Sierra,* paperback edition (Edinburgh: Canongate, 1993).

19 On this approach to Christian ethics, see Harmon L. Smith, *Liturgy and the Moral Life* (Cleveland, OH: Pilgrim Press, 1995). See also Northcott, loc. cit., ch. 7, and Robin Gill, *Churchgoing and Christian Ethics* (Cambridge, UK: Cambridge University Press, 1999), ch. 8.

20 Arne Naess, *Ecology, Community and Lifestyle,* translated by David Rothenburg (Cambridge, UK: Cambridge University Press, 1989).

21 Edward O. Wilson, *Biophilia* (Cambridge, MA: Harvard University Press, 1984).

22 James Nash, *Loving Nature: Ecological Integrity and Christian Responsibility* (Nashville: Abingdon Press, 1991) and Sally McFague, *Super Natural Christians* (London: SCM Press, 1997).

23 This connection between Christian worship and attitudes of respect for nature is not just theoretical. Robin Gill presents empirical evidence that churchgoers are more likely than non-churchgoers to belong to environmental organizations and to be committed to nature conservation and environmental campaigns: Gill, loc. cit., 193–194.

24 Nash, loc. cit., ch. 6.

25 See further Michael S. Northcott, "From Ecological U-topia to Parochial Ecology," *Ecotheology* 9, 2000.

26 Daly, loc. cit.

27 See further Clive Ponting, *Green History of the World: The Environment and the Collapse of Great Civilizations* (London: Penguin, 1994).

28 Stephen Clark, *How to Think About the Earth: Philosophical and Theological Models for Ecology* (London: Mowbray, 1993).

29 See further John Howard Yoder, *The Politics of Jesus* (Grand Rapids, MI: Eerdmans, 1972), ch. 5.

# 4

🌿

# From Ecological Lament to a Sustainable *Oikos*

*Anne M. Clifford*

Over the mountains, break out in cries of lamentation,
over the pasture lands, intone a dirge:
They are scorched, and no one crosses them,
unheard is the bleat of the flock;
birds of the air as well as beasts,
all have fled, and are gone. (Jer. 9:10, NAB)[1]

## Introduction

Throughout the Hebrew Scriptures lament rises from the depths of the human spirit in times of great distress. The lament of ancient Israel is usually focused on the people's suffering due to their infidelity to the sacred covenant that God initiated. What sets Jeremiah's lament apart from many found in the Hebrew Scriptures is its focus on the effects of sinful human choices on the land and the suffering of all creatures dependent on it for existence. The land is ravaged; the people are the cause. The words of Jeremiah herald a "contrast experience." Things ought not to be this way. Something is radically wrong. Human sinfulness has created a serious imbalance in the creation that God has made with "outstretched arm" (Jer. 27:5, NAB).

The lament of Jeremiah, likely proclaimed circa 597 BCE,[2] is limited to a relatively small territory and population. Although the majority of the world's six billion people no longer live in a pastoral society that resembles the Judah of Jeremiah's era, the call to "break out in cries of lamentation" poignantly illustrates the exhortatory power of sacred Scripture to reach beyond the limitation of the time and place

*51*

of its first articulation to us today. As in the case of Jeremiah's situation, we humans have brought devastation upon ourselves and the rest of creation. The difference today is that the devastation is of global proportions. *Oikos,* our earthly home, is imperiled.

## Our present situation and the question of sustainability

Due largely to the indiscriminate application of science and technology, we now inhabit a planet that is under a more dangerous sun, with less arable land, and with a far greater burden of the legacy of poisonous wastes than Jeremiah did. This is part of earth's lamentable story. In 1992 the UN sponsored the Conference on Environment and Development at Rio de Janeiro, popularly known as the "Earth Summit," to address these and other related problems. Economics played a key role in the conversations. Rooted in the Greek word *oikos,* "economics" points to the laws and organization of our planetary household. Those gathered seemed to recognize that economic progress, benefiting a segment of the human population, was not necessarily compatible with the health of earth. This was not a new idea. During the previous quarter of a century, a growing chorus of voices drew attention to the ways in which the human-created economy was unsustainable because it was not compatible with the "great economy" of earth's complex ecosystems.[3]

At the Earth Summit, sustainability emerged as a central focus. Preparations are now underway for a UN World Summit to be held in Johannesburg, South Africa, August 26 to September, 4, 2002. The focus is "sustainable development." Definitions for the term abound. What meaning has the United Nations invested in it? The UN is still using a definition first proposed by the UN report, *Our Common Future* (1987):

Sustainable development is development that meets the needs of the present without compromising the ability of future generations to meet their own needs. It contains two key concepts:

- ☞ the concept of "needs," in particular the essential needs of the world's poor, to which overriding priority should be given; and

- ☞ the idea of limitations imposed by the state of technology and social organizations on the environment's ability to meet present and future needs.[4]

That this definition continues to play an important role in preparations for the 2002 UN World Summit is evident in a recent report made by Nitin Desai, under-secretary general for economic and social affairs of the UN. He applies the "two key concepts" noted above in a manner that sounds very much like a lamentation, however. He bemoans the fact that the record of attaining sustainability since the "Earth Summit" of 1992 is a matter of disappointment.[5] In the past decade, widespread poverty and under-nourishment have not been reduced. No significant progress has been made in improving the conditions of the environment.[6]

The majority of the most influential decisions that impact sustainability are being made in Western countries where Christianity is at least nominally the majority religion. The importance of the sustainability question has been recognized by some Christian churches. Both the U.S. Catholic Bishop's statement, *Renewing the Earth,*[7] and the Presbyterian Church (U.S.A.) document *Hope for a Global Future*[8] address the issue of sustainability. The issue of sustainability is ripe for ecumenical dialogue and ecumenical efforts. "Ecumenism" is yet another term that finds its roots in the word *oikos.* In its broadest sense, ecumenism affirms the belief that the earth is not only one household; it is the household of God. This is usually conceived by Christians in terms of a grand edifice comprised of many churches. But as Christine E. Burke points out, ecumenism also captures "the sense of the whole inhabited earth as one household of God."[9] Mindfulness that the household of God is also the household of life can provide Christians with a common starting point for a theology that realistically acknowledges that there is much to lament about where the health of the earth is concerned. But the mournful dirge of lament can be replaced by hope if we engage our own spiritual traditions in the interest of developing a theological grounding for a sustainable *oikos.*

## A Christian theology of *oikos* and its formidable challenges

Although some churches have published documents that address sustainability, this issue and the related environmental and ecological concerns are in many ways still novel for Christian churches. Churches must respond to these issues not only because of their commitments to the justice teachings of Jesus Christ and the prophets, but also because of their need to defend their own traditions.

The articulation of Christian foundations for a theology of sustainable *oikos* requires that some formidable critiques put before Christian doctrines by environmentalists and ecologists need to be addressed. These critiques have challenged the most basic beliefs that Christians share about creation and its relationship to redemption. By facing these challenges, components for an ecological theology responsive to the question of sustainability can be articulated.

The number of critiques will be limited to two, chosen because they are the most formidable challenges for a Christian theology for a sustainable *oikos*. The first challenge focuses on the Creation texts in Genesis and judges them to be blatantly anti-*oikos*. The second is broader, arguing that Christianity is more interested in having its adherents achieve the goal of "an after-life" (eternal salvation for selected humans) than concern for the earthly household of life.

## Is biblical creation faith anthropocentric at the expense of *oikos*?

This first challenge is not new; it was poignantly raised over three decades ago in a widely cited article written by Lynn White, Jr., a historian of science, to argue that Western Christianity is the major contributing factor in earth's disastrous ecological condition.[10] This statement by White continues to be challenging today: "In its Western form, Christianity is the most anthropocentric religion the world has seen … Man shares, in great measure, God's transcendence of nature."[11] This anthropocentricism, according to White, is evident in the first man, made in God's own image (Gen. 1:26–28), being charged with naming all the animals (Gen. 2:18–19), thus establishing human dominance over them.[12] According to White and many who espouse his reasoning, these divine charges legitimated human exploitation of the earth by the Christian West for human ends. To this challenge, many well-meaning Christians have responded that the Genesis texts in question are really about human stewardship—humans acting as God's representatives on earth. This response, however, leaves many important issues unanswered. Emphasis on stewardship does not sufficiently remedy the anthropocentricism critique, because stewardship presumes that it is the human species that has special status among earth's other life forms, with God-given control of them.

Do Genesis 1:26–28 and 2:18–19 necessarily legitimate an anthropocentric domination of all of nature for human ends, as Lynn White and many after him have claimed? Or have these texts contributed to the ecological crisis due to a tradition of interpretation that is fundamentally faulty? A "yes" to the latter is appropriate if we consider a pattern of interpretation traceable to the dawn of the "Age of Science" and to Francis Bacon. A highly influential philosopher of science of the seventeenth century, Bacon interpreted the Genesis Creation stories as showing that man's control of non-human nature was God's will.[13] In his reading of Genesis 2 and 3, Bacon attributed blame for the first sin on Eve's willful disobedience. Eve, therefore, is responsible for the human race's loss of dominion over the earth. Before the Fall, Adam and Eve, made in God's image, were like God, sharing in God's dominion over earth's creatures, all of which had been given to them by God. With the first sin and subsequent fall from God's favor, this wonderful godly dominion was lost. Bacon envisioned science and its applications in new technologies as the only ways in which man (scientists and those with monetary resources to sponsor their research) can recover the original dominion lost to him through Eve's disobedience and willful temptation of Adam. Since woman's inquisitiveness caused man's fall from his God-given dominion, man's control of another female, "unruly nature," should be used to regain it.

Bacon's interpretation, which weaves together the two Genesis stories of Creation with seventeenth century British gender stereotypes and definitions of progress, uses the biblical texts to legitimate the domination of nature by educated men like himself. His interpretation of the initial chapters of Genesis provided thinking that contributed to the Enlightenment ideas that fed the drive for discovery, the conquest of continents and the exploitation of nature in distant lands around the globe. With its emphasis on the autonomous individual (i.e., the white male of financial means), the Enlightenment's guiding principles contributed to a structured anthropocentricism that envisioned progress in terms of the extension of the power of the educated moneyed class in the achievement of industrial ambitions. The result was that many wealthy people gave very little priority to the common good of human communities and no priority to non-human life, which was virtually reduced to a mere instrumental means to the achievement of specific economic goals.[14]

As for the troublesome texts that White highlighted and Bacon interpreted to provide religious legitimacy for the work of scientists and the developers of technology, they can be best understood if we give attention to their respective historical contexts and look for clues into their more probable meanings. Since we have no direct access to the minds of the biblical authors, the clues will provide us with material for an imaginative reconstruction of the "Days of Creation" and the "Creation of Adam and Eve" stories.

The second account of Creation, as found in Genesis 2, will be treated first because it was written much earlier, roughly five centuries before the first chapter of Genesis. It will also be treated first because the argument that the directive to Adam to name the animals is permission to dominate them is easy to dismiss. Likely written during the time of David and his successor Solomon (1010–930 BC), the ancient story in Genesis 2 is attributed to the Yahwist source. As the tale unfolds, Yahweh-God is the dominant character. Yahweh forms the first human from the dirt of the earth (v. 7). The gift of life-breath that God breathed into the first human makes this creature a living being, whom God provides with the means to sustain life. God places this first human in the Garden of Eden, a place of plenty that the human earthling (sexual differentiation awaits the creation of the first woman) is to cultivate (v. 15). The story progresses with God recognizing that it is not good for the earth creature to be alone and making a suitable partner. From the same ground used to form the earth creature, God forms beasts and birds and brings them to the man to see what he would call them. The man gives names to all the animals (Gen. 2: 18–19, NAB). Naming in this context is not a divine directive for humans to exploit animals for human ends. By naming, humans relate to the thing named, whether plant, animal or inanimate object, in a way unique to our species.[15] Naming does not change the fact that humans and the beasts and birds are all created from the substance of the earth and are bound to it.

The account of the "Days of Creation," in contrast to the so-called "Story of the Creation of Adam and Eve," is far less explicitly earthy. There is a broad consensus among biblical scholars that Genesis 1:1–2:4a was composed during or shortly after the era of the Babylonian exile (circa 587–540 BCE). This was a time of crisis for Jews enslaved in a foreign land (present day Iraq). In the midst of considerable uncertainty, the authors of this tradition, a part of the priestly writings likely written for use in public prayer, reasserted their belief in God's power to order chaos in a creation narrative. This narrative clearly includes among its intentions the praise of God who provides a Sabbath day of rest for all of creation for this purpose. What could be dearer to slaves exiled in a foreign land than to once again be able to freely praise their God?

The priestly Creation narrative unfolds with God creating humans on the same day on which God also created animals. The text explains that God said:

> "Let us make man in our image, after our likeness. Let them have dominion over the fish of the sea, the birds of the air, and the cattle, and over all the wild animals and all the creatures that crawl on the ground." God created man in his image; in the divine image he created him; male and female he created them. God blessed them, saying: "Be fertile and multiply; fill the earth and subdue it. Have dominion over the fish of the sea, the birds of the air, and all the living things that move on the earth. (Gen. 1:26–28, NAB)

Since this story has humans sharing the same day of Creation with the animals, it would seem that there is a natural kinship of humankind with all animal species. This kinship, affirmed differently in the common earthiness of Genesis 2, is underscored when humans are told by God that they may not eat animals; plants alone will provide them with food (Gen 1:29). This important verse provides a perspective from which "subdue" (Heb. *kavash*) and "dominion" (Heb. *yarad*) should rightly be interpreted.

Of the so-called "troublesome directives," the more vexing is "to subdue the earth." Modern English speakers might associate the word "subdue" with control and subjugation. On the surface, subduing the earth may seem to promote human power over and exploitation of the earth and its life forms with negative outcomes. Perhaps more insight can be provided for the biblical meaning of "subdue" if we examine how it is used elsewhere in sacred Scripture. The Hebrew word for subdue appears in Numbers and 1 Chronicles. In Numbers 32, subdue is synonymous with what is required for the Jews to inhabit, with God's help, the land promised to them.[16] In 1 Chronicles 22 "subdue" is again found with reference to land that is to be prepared for the construction of the temple, a "sanctuary of the Lord."[17] One might reasonably conclude, therefore, that "subdue" conveys the meaning of transforming land into a home for the people of Israel, a land where God can also be worshiped as the covenant prescribed.

What bearing does this have on Genesis 1:28? In light of the heartbreaking loss to the Jews of their homeland and temple, and their enslavement as exiles in Babylon, "subdue the earth" may simply mean that the Jews who were returning to their former home believed they had a responsibility to have children and rebuild a society on their former land. The biblical directive to subdue the earth cannot be simplistically equated with a license to exploit non-human creation, as if it were a mere instrument for human use. It is better interpreted as a directive to reclaim a divine gift, the "Promised Land," where the covenant could be kept and God suitably worshiped.

Human dominion over creatures found in Genesis 1:26 and 28 is repeated elsewhere in the Bible[18] but, according to Richard Clifford, there are good reasons for arguing that the fuller meaning of "dominion" emerges in Genesis 6 to 9.[19] In Genesis 6, land, animals and people are not said to be lamenting; rather, it is God who laments over "how great was man's wickedness on earth" (Gen. 6:5, NAB). Their sins will bring upon the earth an ecological disaster of global proportions—the "Great Flood." The story focuses on Noah and his family, who alone are righteous in the sight of God. God instructs Noah to build a huge ark and directs him: "Of all kinds of birds, of all kinds of beasts . . . two of each shall come into the ark . . . to stay alive" (Gen.6:20, NAB). God's directive gives content to human dominion: Noah and his family are charged with seeing to the survival of the other living creatures. Dominion exercised by God's righteous ones results in "salvation"—the deliverance of animals from destruction and, after the Flood, the flourishing of humans and every kind of bird and beast. Animal and human survival are intimately related.

The climax of the story is a special covenant that God initiates with Noah's family and their descendants, and, very importantly, with every living creature—"all the birds, and the various tame and wild animals" (Gen. 9:9–11, NAB). This new covenant underscores the inherent relational interconnection of humans with the rest of creation. The Noahic covenant is a symbol of the unbreakable bond of all creatures with their Creator and with one another. The perpetual sign of this covenant is God's "bow in the clouds" (Gen. 9:13, NAB), a reminder of the relationship God has not only with humans but also with all living creatures.

While the authors of these ancient Genesis texts were not concerned with ecological disaster resulting from extensive exploitation of non-human nature, they are relevant for a theological understanding of sustainability. This understanding honors and supports humanity's kinship solidarity with all of the other life forms—birds and beasts and "all kinds of creeping things of the earth" (Gen. 1:25, NAB)—and honors that kinship solidarity as a sacred covenant. Humans have a moral responsibility to sustain the order of the world God created.

## Does biblical faith in redemption put *oikos* in the background?

In response to the second challenge, it is true that Christianity places great emphasis on the redemption offered through Jesus Christ and human salvation. Beginning in the Enlightenment period, this emphasis contributed to the neglect of non-human nature and to ecological disasters. However, a careful look at the biblical sources shows that the neglect of *oikos* overlooks the ways in which God's work of creation provides the cosmic purpose behind God's redemptive activity. This is true for the Old Testament[20] as well as for the New. In Psalm 146 (NAB), for example, we find a hymn composed by someone who has learned that there is no other source of salvation (v. 3), than God the Creator, "the maker of heaven and earth, the seas and all that is in them"(v. 6).

It is this same God who

> secures justice for the oppressed,
> gives food to the hungry…
> sets prisoners free…
> gives sight to the blind…
> raises up those who are bowed down…
> protects the stranger…
> [and] sustains the orphan and the widow,
> but thwarts the way of the wicked. (vv. 7–9)[21]

In each activity, God responds to the creatures most in need, offering them the liberation of a redeemed life.

In the New Testament, creation and redemption are treated as two related aspects of God's one engagement with the world in and through Jesus Christ. Through the incarnation, God comes into relation with the world of creatures in a personal and intimate way. In a manner that resonates with Psalm 146, Jesus brings glad tidings to the poor, proclaims liberty to prisoners, gives sight to the blind, and secures justice for the oppressed (Luke 4:18). While it is true that these

activities are directed towards people, this passage ends with a proc-
lamation of a Sabbath year. This is a year of favor in which not only
are slaves to be freed and debts cancelled, but also, planting, pruning
and harvesting for storage are forbidden. The earth itself is to be given
Sabbath rest in honor of the Creator (Lev. 25: 2–7, Deut. 15:7–11).

Further, through Jesus' life, death and resurrection, God's creative
activity continues as a work of redemption. This is clearly affirmed in
an early Christian hymn in Colossians that proclaims that God,
through Christ, the first born of all creation, reconciled to himself all
things (human and non-human creatures), whether on earth or in
heaven (Col. 1:1–20). The saving work of Jesus Christ, therefore, cannot be simplistically limited to "an after-life" for human beings. Redemption in Jesus Christ is not reductively anthropocentric. It extends
to the entire household of life with God embracing all creatures in
and through Jesus Christ.[22]

## Conclusions

This limited treatment of creation and redemption provides a theological basis for approaching sustainability in an explicitly ecological
way. The definition of sustainable development adopted by the United
Nations places emphasis almost exclusively on human economic development. It is laudable for its concern for the poor whose very existence is threatened daily, as the gap between the economically poor
and the affluent continues to widen. Yet the definition is myopically
anthropocentric in its emphasis. It supports concern for the environment in so far as it must be sufficiently healthy to meet present and
future human needs. It contains a very important emphasis, but it
does not bring a holistic ecological consciousness to bear on the world
situation. It fails to acknowledge that humans are but participants in
a highly complex network of life comprised of delicate ecosystems.

In response to the two major challenges to Christian faith addressed above, a kinship solidarity of humans with all forms of life
and with the earth itself emerged. It is affirmed in the Genesis Creation texts, symbolized by the Noahic covenant God initiated for the
"common good" of all living creatures, and subsumed in the saving
work of Jesus Christ that envisions redemption encompassing human
and non-human creation. The working definition of sustainable development employed by the United Nations does not encompass the
biblically rooted kinship solidarity presented here.

An arguably more adequate definition of sustainability is that of Larry Rasmussen, who speaks of sustainability as "the capacity of natural and social systems to survive and thrive together indefinitely. It is also a vision with an implicit earth ethic . . . and a picture of earth as *oikos*."[23] This definition is preferred because it takes into account the inherent capacity of earth systems to strive to maintain their own balance and replenishment. It also places humanity within the ecosphere as a participant, rather than as the sole referent in determining global policies. It further recognizes that an earth ethic must be oriented to *oikos* as a whole, and therefore be thoroughly ecological.

An essential component for a Christian theology of sustainability, conscious of kinship solidarity, is the recognition of the close link between both human poverty and affluence, and the degradation of the health of *oikos*.[24] Socioeconomic injustice among humans is evident in statistics that indicate that 1.2 billion of the 6.1 billion people on the planet are overweight, while an another 1.2 billion are seriously malnourished.[25] The affluent of the Northern hemisphere, where most of the 1.2 billion obese reside, consume the majority of the earth's resources and create most of its non-biodegradable and toxic wastes, resulting in species extinction that is wiping out forms of life that took millions of years to evolve. *Oikos* can no longer afford the minority affluent who define well-being in terms of their own accumulation of wealth and resources.

Kinship solidarity, foundational for a Christian ecological theology of sustainability, provides us with a "world vision" that extends the realm of justice to the whole of creation. It takes seriously that *oikos* is not only the household of life, but also the household of God. As the household of God, *oikos* has an inherent sacrality even a sacramentality.[26] Every creature is engraved with the unmistakable marks of God's glory.[27] For Jeremiah's lament to be replaced by an *oikos* rejoicing with cries of gladness (Ps. 100:1–2), a program for a sustainable eco-justice—one which affirms the intrinsic value of all creatures—must be adopted. Deep-seated anthropocentric attitudes and practices that have long gone unchallenged must be replaced with an emphasis on the integrity of all of God's creatures if a sustainable *oikos* is to become a reality.

---

[1]  The selected verse is from a longer section, Jeremiah 8:4–10:25, in which the dominant theme is disaster and the need for repentance. Lamentation is also the central

theme of Jeremiah 14:1–15:9, where drought and war, famine and sword are inter-woven. See also Hosea 4:3 ("Therefore the land mourns, and everything that dwells in it languishes: The beasts of the field, the birds of the air, and even the fish of the sea perish" [NAB].) and Isaiah 24:4–5 ("The earth mourns and fades, the world lan-guishes and fades; both heaven and earth languish. The earth is polluted because of its inhabitants, who have transgressed laws, violated statutes, broken the ancient covenant" [NAB].).

2  The occasion of this lament is likely Nebuchadnezzar's first campaign against Judah in 597 B.C.E. See Guy P. Couturier, C.S.C., "Jeremiah," in *The New Jerome Biblical Commentary*, eds. Raymond E. Brown, S.S.; Joseph A. Fitzmeyer, S.J.; and Roland E. Murphy, O. Carm.; (Englewood Cliffs, NJ: Prentice Hall, 1990), 276.

3  Wendell Berry, *Home Economics* (San Francisco: North Point Press, 1987). Berry ar-gued that the human economy must be "an analogue of the Great Economy" in how humans use earth's limited resources or the ecological crisis would continue to esca-late, 59.

4  The World Commission on Environment and Development chaired by Gro Harlem Brundtland, *Our Common Future* (New York: Oxford University Press, 1987), 43.

5  Nitin Desai, "Statement to the Second Committee, Introducing Item 98: Environment and Sustainable Development" <http://www.johannesburgsummit.org/> 29 Octo-ber 2001.

6  Ibid. 3 and passim.

7  Sustainability appears 16 times in "Renewing the Earth, an Invitation to Reflection and Action on Environment in the Light of Catholic Social Teaching" in *And God Saw that It Was Good: Catholic Theology and the Environment*, eds. Drew S.J. Christiansen and Walter Glazier, (Washington, DC: United States Catholic Confer-ence, 1996), 223–243.

8  *Hope for a Global Future: Toward Just and Sustainable Human Development* (Louis-ville: The Office of the General Assembly, Presbyterian Church, USA, 1996), 76.

9  Christine E. Burke, I.B.V.M., "Globalization and Ecology," in *Earth Revealing, Earth Healing* (Collegeville, MN: Liturgical Press, 2001), 21.

10 Lynn White, Jr., "The Historical Roots of our Ecologic Crisis," first published in *Sci-ence* 155 (1967), 1203-07; reprinted in *Readings in Ecology and Feminist Theology*, eds. Mary Heather MacKinnon and Moni McIntyre, eds. (Kansas City, MO: Sheed and Ward, 1995), 25-35.

11 Ibid., 31. Interestingly enough, White called for a renewed Franciscan spirituality, because Francis "tried to substitute the idea of the equality of all creatures, including man, for the idea of man's limitless rule of creatures," 35.

12 Ibid., 30-31. Later White writes: "We believe ourselves to be superior to nature, even contemptuous of it, willing to use it for our slightest whim," 33.

13 See Carolyn Merchant, *The Death of Nature: Women, Ecology and the Scientific Revo-lution* (San Francisco: Harper and Row, 1980), 170. Merchant cites "Novum Or-ganum," Part 2 in Francis Bacon, *Works*, 4: 247 and "Valerius Terminus," in *Works*, eds. James Spedding, Robert Leslie Ellis, and Douglas Heath, (London: Longmanns Green, 1870), 3: 217, 219 and "The Masculine Birth of Time," in *The Philosophy of Francis Bacon* ed. Benjamin Farrington, (Liverpool, England: Liverpool University Press, 1964), 62, fn. 13, 317.

14 This argument is obviously one-sided if the positive contributions of the Enlighten-

ment are not also acknowledged: science ridding the world of many devastating diseases, recognition of human rights, democracy replacing tyranny, and tolerance for religious differences are among the most significant.

[15] See Claus Westermann, *Genesis 1–11, A Commentary*, translated by John J. Scullion, S.J., (Minneapolis: Augsburg Publishing House, 1974), 228–229.

[16] For more on "subdue," see Richard J. Clifford, S.J., "Genesis 1–3: Permission to Exploit Nature?" in *Bible Today* (1988): 135. In Numbers 32:21–22 and 29–34, "subdue" is used in reference to Moses giving directives to the priest Eleazar, to Joshua, and to the heads of the ancestral tribes of the Israelites to cross the Jordan, enter Canaan, and subdue the land, building towns on it.

[17] In 1 Chronicles 22:18–19 the land occupied under David is subdued and the people are directed: "devote your hearts and souls to seeking the LORD your God. Proceed to build the sanctuary of the LORD God, that the Ark of the covenant of the LORD and God's sacred vessels may be brought into the house built in honor of the LORD" (1 Chron. 22:19, NAB).

[18] Among them are Psalm 8:6 and the Deuterocanonical/Apocryphal texts: Wisdom 9:2 and Sirach 17:4.

[19] Richard Clifford, loc. cit., 136.

[20] For more development of this point, see Terence E. Fretheim, *Exodus: Interpretation, A Bible Commentary for Teaching and Preaching* (Louisville, KY: John Knox Press, 1991), 12–14, passim.

[21] For another Psalm that treats the themes of creation and redemption, see Psalm 19.

[22] In Jesus Christ, the transcendent and the immanent, the "other-worldly" and the "this-worldly" meet.

[23] Larry Rasmussen, *Earth Community, Earth Ethics* (Maryknoll, N.Y.: Orbis Books, 1996), 127.

[24] Ibid., 103.

[25] Data are from the World Health Organization, which also estimates that an additional 2 billion are on the edge of hunger on a nearly daily basis, cited by Gary Gardner and Brian Halweil, "Underfed and Overfed: The Global Epidemic of Malnutrition," *WorldWatch Paper* 150, March 2000, 7.

[26] The sacrality and sacramentality of creation is a Roman Catholic insight traceable to the writings of Thomas Aquinas, who envisioned creatures participating in the divine goodness and manifesting vestiges of the Trinity; see the *Summa Theologiae* I, questions 47, a. 1 and 45, a. 7.

[27] This is a paraphrase of a statement made by John Calvin, *Institutes of the Christian Religion*, I, v.1.

# 5

 ❧

# Stealing Creation's Blessings

*Don Brandt*

## Introduction

Spearheaded by global warming, environmental issues are back in the news and are likely to reach a crescendo by August 2002 during the UN Conference on Sustainable Development. Many opinions from dozens of groups will be expressed at this conference. Clear, prophetic Christian voices speaking out on today's twin problems of environmental complacency and unrestrained exploitation are needed to clarify and heal divisions among people, communities and nations. Unfortunately, healers—including Christian environmentalists—are besieged by opposing forces both within and outside of the church.

The main thesis of this chapter is that God commanded humankind to care for the earth. Failure to do so is an affront to the Creator. Nowhere does the Bible speak of anything less than an attitude of gratitude, justice and love towards creation. Evangelism does not compete with this way of thinking; it flows out of it. The fact is that all of us are guardians over pieces of creation. We care for our bodies, protect and educate our children, have life, health and other insurance policies in wealthier countries, repair the car, maintain the house, and weed the yard. We need to transfer that attitude of creation concern to the biosphere, our environment, that's home to all of us. Christian naturalist and thinker Hans Schwarz said it well: "The failure to conserve the land is not only a misuse of God's gift but a sin."[1]

Mainline or conciliar Protestant churches have taken a stand for the environment, as have Roman Catholic and Orthodox churches. Over the past two decades most evangelical congregations have joined them. In theory at least, most Christians believe in a triune God who has given humankind sacred responsibility for the earth and all its creatures. Yet some forces within the church rail against environmentalism. By expressing an ardor for the bios, Christian environmentalists are seen as borderline pagans embracing pantheism. The anti-environmental camp points to Christians who have leaped the bounds of orthodoxy and encamped with the eco-feminist, Gaian, and other New Age "enemies," proving that to be an environmentalist is courting the Devil. The worldview of other Christians is that "we are just passing through" and the "end times" are near. Why bother with snail darters and condors when there are millions of people to save and there is so little time to do it? Environmental concerns pale to irrelevancy in the minds of these fervent evangelicals.

The church has made it easy for some secularists to blame Christianity for the environmental mess in which the world finds itself. Believing the church is an obstacle to ecology, many scientists put their faith in technology and technical environmental fixes. Others, such as many eco-feminists and deep ecologists, seek to change social structures that they say harm the environment. Still others view the environmental crisis fundamentally as a religious issue. Turned off by Christianity, they turn to other religions including earth worship.

## Trinitarian perspective

Belief in a triune God who is Creator, Redeemer, and renewer means that Christian environmentalists are unabashedly theocentric in outlook. The human-centered or anthropocentric label often given to Christians—that of domineering utilitarian managers of the environment—is not biblically accurate. God is Creator of the cosmos. Humans and the rest of the universe, whether animate or inanimate, are the created. As a created and dependent life species, people are part of creation. This means that we are in the same class (the created) as animals, plants, air, water and rocks. However, humans have a special role among the created in reflecting the image of God. One principal way we do this is to obey and glorify God in our care for the rest of creation.

God is Redeemer. Redemption is seen in the life and work of Jesus Christ who has, is, and will restore all things (Colossians). It is Christ who heals a world separated from God and broken by sin. God the Holy Spirit is recreating and renewing our lives and the universe. We experience this at the cosmic level as astronomers continually discover new marvels in the universe. We can also view God's wonders at the micro level as biochemists continually fascinate us with discoveries in the working of cells, DNA, and genes.

### Stewards

As stated in Genesis, "dominion" means that humans are to care for creation. This is seen in Genesis, where God gave humankind the privilege of naming created species. Naming brings with it a powerful tie between both parties. The garden image in Genesis connotes that humans are caregivers and caretakers of God's creation. Most Christian environmentalists use the term "steward" to convey the idea that we are part of the environment (the created) while having special responsibilities as worshipful custodians.

This role of Christians as environmental stewards is ignored, criticized, or shunned by many groups, both Christian and secular. Some people are unaware of how their lifestyles negatively affect the environment. Others simply don't care. In the developed world, avarice feeds on consumerism, producing hundreds of environmentally harmful products that meet few real needs. Forests are ravaged for timber. "Exotic" birds, fishes and other animals are transported from the tropical South to the industrial North. Poverty plays its part, too, in extraction and depletion of species and general environmental degradation. The poor do cut trees and deforest land for firewood and building materials. They also farm steep slopes and thereby accelerate soil erosion. The rural poor are not ignorant about the effects of their actions. They know that what they do is harmful to the land, but survival gives them little choice. Structural changes that bring land, market and credit reforms can help create environmentally sustainable farming practices.

## Christian critics

The views of too many Christians towards caring for the environment range between "why bother?" to outright hostility. Generally, these are evangelicals who strongly believe that the world as we know it will soon pass with the impending coming of Christ. If this is one's view, it follows that priority must be placed on converting as many people as possible before Christ returns. Energy placed in other efforts, such as "saving" the environment instead of souls, is frivolous and irrelevant. It may even be the work of Satan, as resources are drawn away from world evangelization.

Fault cannot be laid at the door of those eager to fill the Great Commission (Matt. 28:19–20). We may question the scriptural basis that the world has entered the end times era[2] even as we empathize with these believers. Much that we see and hear certainly isn't nice: rampant alcohol and drug abuse, increased domestic and street violence, trafficking of women and girls in epidemic proportions, court-protected pornography, the persistence of some diseases and the comeback of others. Yet we may question if the Lord excuses us from social and environmental responsibilities and demands us to wholly focus on evangelism by the spoken word. In writing on this issue, it is a major concern of Ronald Sider[3] that people are repelled by the gospel message because they see Christian indifference and carelessness towards the rest of creation.[4]

### Christian negligence

This raises broader questions. Why do many Christians—and not only those who believe the end times are near—have a cavalier attitude towards the environment? Why don't Christians take the command to care for the environment seriously? Is it due to a lack of knowledge and discipling? Are most Western Christians dualists who do not see themselves as part of creation? Is our faith in technology and the market so strong that we think they can somehow solve all problems? Or is it because our baals—the gods of consumerism and greed—are too strong to resist?[5]

Perhaps no Christian environmentalist is better known and respected than Calvin DeWitt. He relates that the road to environmental care and nurture involves a three-step process.[6] The first step involves people becoming aware of creation. We need to identify and name God's creatures. Naming is a way of getting better acquainted and establishing a relationship with the environment. For the ancient Hebrews, naming was considered a moral responsibility and a sign of respect for life.[7] The second step is appreciation. People need to tolerate, respect, and value creation. This may be a long educational process that requires hands-on experience. The hardest task may be to motivate "couch potatoes" to walk in a park using all their senses. Appreciation leads to concern of the environment. This is the third step, in which the "I-thou" dualistic mind-set of humans versus nature becomes "us," and stewardship is born. We realize that we are part of creation. Attitudes we once held that humankind were separated, and therefore "better" than the rest of creation, have disappeared.

## The Church

The generally tepid attitude of Christians towards the environment is disturbing. This is especially so because most Christian denominations and persuasions take the position that care for the cosmos is the responsibility God gave to humankind. Today's Roman Catholic theologians, for example, recognize how the early church was influenced by neo-platonic belief that emphasized the separation of soul and matter. This dualism led to humans juxtaposed against nature and is expressed in the writing of such pre-eminent theologians as St Augustine (354–430).

While dualism predominated, more holistic views have a long "counter-cultural" history. Recognition that humans are part of the created order is found in the writings of the early church. Unlike Greek philosophers, Irenaeus (d. 208) thought that the material world was fundamentally good.[8] Similarly, the poetic earthiness of the Celtic fathers shows a strong bond between humankind and the rest of nature and nature with the Creator. This immanence—seeing God in nature—continued through the centuries and is clearly revealed in the early twentieth century verse of Irish poet Joseph Plunkett:

> I see His blood upon the rose,
> And in the stars the glory of His eyes,
> His body gleams amid eternal snows,
> His tears fall from the skies.
>
> I see His face in every flower,
> The thunder and the singing of the birds
> Are but His voice—carven by His power
> Rocks are His written word.
> All pathways by His feet are worn,
> His strong heart stirs the ever-beating sea,
> His crown of thorns is twined with every thorn,
> His cross in every tree.[9]

### Roman Catholic

Catholic environmental theology is rooted in the teachings of St Thomas Aquinas (1224–1274) and St Francis of Assisi (1181–1226). Aquinas regarded humans as part of the created world. Along with the rest of creation, people are clearly separated from the Creator. The environment is both plentiful and inherently good:

God deliberately brings about multitude and distinction in order that the divine goodness may be brought forth and shared in many measures. There is beauty in the very diversity." [10]

At the same time, St Thomas has no doubts about the uniqueness of humans in the rest of creation. Humans are the highest in the hierarchy of the created order, with special rights and responsibilities. Perhaps humankind's role may be best likened to that of a corporate manager or CEO of God's creation—not a dominating dictator, as the Lord remains the ultimate "boss," but not a steward either. Still, Aquinas believed that knowledge of God could be acquired from reading Scripture and reflecting on the cosmos (natural theology). Both may serve as guides to moral living as the conscience is a "natural voice" and part of "natural law."

Until fairly recently, the image of steward or earth-keeper was relegated to a minority of theologians, of whom St Francis of Assisi is the best known. Statues and pictures of St Francis feeding birds or petting wild animals abound. They are fixtures in homes, churches, parks, and stationery. One is tempted to say that among most environmentally sensitive Christians, St Francis has become something of a fetish. He is the "patron saint" of both Roman Catholic and Protestant environmentalists. St Francis' love for creation was truly radical and seems to be based on the passion that he had for everything in God's universe. Francis was indeed the first "deep ecologist," but like Aquinas, his views of creation and the bios were theocentric and Trinitarian.

Today, Roman Catholic environmental theology is very close to Orthodox theology and generally in harmony with mainline and most evangelical Protestants. Few Christian ecologists would differ with Clifford when he states that "there is an inherent relational interdependence of humans with the rest of creation and their creator."[11] Differences with many Protestants may lie with sacramentizing nature. This may be a semantic issue with some Protestant groups; with others it's a theological one. God transcends (is above) creation, but God is also seen in nature (immanence). This means that the environment can bless humans by bestowing God's grace on us. As stated by the U.S. Catholic Bishops, sacramentizing means a " . . .world that discloses the Creator's presence by visible and tangible signs" (i.e., nature).[12]

Most Protestants have no trouble in thinking that nature can inspire people with thoughts of God's power and majesty or that the cosmos is truly God's handiwork. To move a theological step further and believe that God both transcends the created and may actually be found in nature may be beyond their ken. Instead of affirmation, it's more likely to spark protests that immanence smacks of New Age pantheism. Yet Scripture says that God blessed Creation and called it good. After the Fall, God is repeatedly seen in nature (1 Kings 19:12) and nature is seen responding to the Creator (Ps. 96:12). Is it not a logical step from here to see how nature can be a blessing to humans, and humans to the rest of creation?

Another difference with the views of some Protestants is that Catholic teaching directly links stewardship of the environment with care of the poor. John Paul II made this tie in his 1990 World Day of Peace message. The U.S. Catholic Bishops developed the theme further in their 1991 pastoral paper, "Renewing the Earth." The basic issue is biblical justice. Catholic social teaching promotes the inalienable dignity of every person. Minimizing environmental damage will benefit everyone, especially the poor. The poor, much more than other income groups, pay the price for environmental desecration. The urban poor in the North live closest to oil refineries, chemical dumps and electric power stations. In the South, the location of shantytowns correlates with putrid, flood-prone streams or hills subject to mudflows. Justice for the land translates into justice for the poor. As stated by Christensen and Grazer, "protection of the earth and the human community go hand in hand . . . a truly classic case of the common good."[13]

### Orthodox

Similar to Roman Catholic theology, Orthodox theology includes the tie between the poor and the environment. Orthodoxy, too, has a rich legacy of thought that interprets "dominion" and "subjection" as humankind's responses to be creation's caretakers. This is seen in a couple of ways. One is the call to protect the natural environment and cherish the biodiversity of the planet.[14] A second is the deep-rooted belief in a "single" creator and creation propounded by the seventh century mystic, St. Isaac of Syria.[15]

Another characteristic of the Orthodox worldview is the sacramentizing of creation that follows the firm belief that God is found in nature. In the emphatic and elegant words of nineteenth century novelist Fyodor Dostoyevsky,

> Love all of God's creation, the whole of it and every grain
> of sand. Love every leaf, every ray of God's light! Love the
> animals, love the plants, love everything. If you love ev-
> erything, you will soon perceive the divine mystery in
> things. Once you perceive it, you will begin to compre-
> hend it better every day. And you will come at last to love
> the whole world with an all-embracing love.[16]

This immanence quality of God in the environment may be more
strongly held in the Orthodox than the Roman Catholic faith. This
could be due to the smaller influence of neo-platonic dualism in Orthodoxy. At the same time, Western Enlightenment, with its emphasis on individual utilitarian pursuits, didn't penetrate many Eastern
lands. [17]

According to Orthodox theology, there is unity between man and
creation only when creation is used in accordance with nature. Only
such a use guarantees that the inner principles which the Maker has
placed in things are saved and preserved. This is not only a commandment of God, but also an essential precondition for preserving the
balance of nature itself.[18]

A more subtle difference between Orthodox and Roman Catholi-
cism is the ecological view of the Last Supper. Both faiths place the
Eucharist in the centre of corporate worship. In the Orthodox tradition, bread and wine remind those who participate in the sacrament
that the material world, and not only humankind, is worthy of redemption.[19] The Eucharist serves as a continual sign that Jesus died for all
of creation and that God's purpose for the cosmos is harmony and
order through the atoning blood of Christ.

### Protestant

Not surprisingly, Protestant positions on the environment are more
diverse than Roman Catholic and Orthodox beliefs. If there is a common thread among Protestants, it is the foundational importance of
the Bible to theological discourse. Within Protestantism, mainline or
conciliatory churches have adopted strong ecological positions since
the 1970s. This is reflected in the actions of the World Council of
Churches and affiliated national councils. How much of these environmental concerns are filtered down to parishioners through sermons, church school lessons, and other means of instruction is un-
clear.

Much of the Protestant evangelical church leadership adopted the concept of environmental stewardship in the 1980s.[20] As in other movements, some parts of the evangelical community are in the forefront and serve as mentors, prophets, and guides. Examples include Evangelicals for Social Action, Au Sable Institute, A Rocha, and the International Evangelical Environmental Network.[21] The case for stewarding the earth has so penetrated evangelical theology that environmental care articles are found in the leading evangelical magazine, *Christianity Today*, on a regular basis.[22]

The seriousness with which evangelicals view their role as stewards may be seen in the 1994 document, "An Evangelical Declaration on the Care of Creation" (Appendix 1). This statement is an elegantly and strongly worded affirmation that biblical faith is essential to the amelioration of environmental problems. Consistent with Orthodox, Roman Catholic and conciliar or mainline Protestant thinking, crass ecological damage is considered a sin. The triune God is affirmed as transcendent Creator who is intimately involved with his creation. This image of the immanence of God also moves evangelists closer to the ecological position of other Christian persuasions. Generally more than conciliary Protestants' proclamations, though, the Evangelical Declaration emphasizes Christ as healer-redeemer who restores wholeness to humankind and all the created order.

Sympathetic critics point to two omissions in the Evangelical Declaration. Sider's concern is that the word "alone" was dropped from the biblical teaching that "people *alone* have been created in the image of God"[23]: "So God created humankind in his image, in the image of God he created them; male and female he created them" (Gen. 1:27, NRSV). Humankind's image of God gives us a unique responsibility among the created. Only people can serve as God's stewards.

Michael Northcott raises a second concern. Although emphasis is correctly placed on humankind's role as stewards, not enough attention is given to worshiping God. This is so since

> worship, true or false, is at the heart of our ecological crisis. It is precisely the modern devotion to the cult of consumerism which is driving the horrific global scale of environmental destruction.[24]

Although varied, Protestant thought about the role of humans in the environment has narrowed into something resembling a broad consensus under the stewardship theme. Differences are now expressed in how much a "caregiver" of nature versus an "overseer" of the environment humankind should be. The role of steward also generally brings Protestant thinking about the cosmos into closer alignment with Orthodox and Roman Catholic perspectives. Yet it's fair to say that a common weakness in the West is that stewardship is not done in community, "where gifts of caring and community can be freely and thankfully shared."[25]

While most Protestants don't sacramentalize the created order, more are seeing a Creator that is both transcendent and immanent. The Bible continually and dramatically shows God's transcendence over the cosmos.

> Where were you when I laid the foundation of the earth?
>    Tell me if you have understanding.
> Who determined its measurements—surely you know!
>    Or who stretched the line upon it?
> On what were its bases sunk,
>    or who laid its cornerstone
> when the morning stars sang together
>    and all the heavenly beings shouted for joy?
> (Job 38:4–7, NRSV)

Evangelicals often overlook God's immanence in nature although it's clearly revealed in the Bible. Without immanence, God's transcendence is deism. Immanence without transcendence leads to panentheism,[26]

> for in him [Christ] all things in heaven and on earth were created, things visible and invisible, whether thrones or dominions or rulers or powers—all things have been created through him and for him. (Col. 1:16, NRSV)

> Ever since the creation of the world his [Christ's] eternal power and divine nature, invisible though they are, have been understood and seen through the things he has made. (Rom. 1:20, NRSV)

Our all-powerful transcendent and immanent God demands to be worshiped. The role of a worshipful steward is readily seen in Old Testament passages. The opening chapters of Genesis reveal that before the Fall, God called all creation very good. The Creator told humankind to care for creation (Gen. 2:15). "Dominion" (Gen. 1:26, NRSV), subjugation (Gen. 1:28) and naming (Gen. 2:20) are creation-caring tools to bring harmony and order to the cosmos. Domination and subjection that conjure up images of humans desecrating the environment by acts of ruthless abandon are contradictory to the Genesis story and the teachings in the rest of the Bible. It makes no sense for God to call something "very good" only to allow it to be abused and squandered.

Not only in Genesis 1 and 2 do we find God's love for creation. That driving concern is reflected later in Genesis following the Flood, as Clifford reflects in her chapter. God's covenant is not only to Noah, and through him all humankind, but to all creation:

> Behold I establish my covenant with you and your descendants after you, and with every living creature that is with you, the birds, the cattle, and every beast of the earth with you, as many as came out of the ark. (Gen 9:9–10, RSV)

Despite the emphasis given to Genesis, perhaps the best way to see God's love for the biosphere and humankind's role as steward is through the Sabbath laws. God demands that not only man but the beasts under his care must rest every seventh day (Lev. 25:1–12). Perhaps even more striking is the law that every seventh year the tilled earth should remain fallow. Fallow fields help restore fertility, guaranteeing long-term productivity. Yet the main reason for the land to rest has more to do with care of the land as a living organism than human ability to grow more crops. A good steward cares for the land. Unfortunately, good stewardship is not practiced by many of the world's farmers, herders and foresters today.

The result is a loss of biodiversity. As created beings, humans are to live in cooperation and harmony with the rest of creation. Deep ecologists and eco-feminists are right when they claim that survival of the species depends more on cooperation than competition. Cooperation implies that there must be a variety of created things, otherwise there would be no need to live together in a reasonably harmonious fashion. Cooperation also implies that variety must be good, or lead to something that is mutually satisfying. In other words, variety is inherently beneficial to the created order. That being so, perhaps the greatest environmental harm that humans are perpetuating is the rapid eradication of other species in the created world.[27]

Currently, more than 11,000 plant and animal species are threatened with extinction. This number includes 24 percent of mammals and 22 percent of bird species. The main culprit is lost habitat through human activities. Many scientists agree with the eminent U.S. scientist E.O. Wilson that the planet is in the early stage of "massive extinction of ecosystems and species."[28] Clearly, the UN Convention on Biological Diversity signed by 175 countries is largely ignored.

Human behavior has an adverse effect on other species. This doesn't mean that people should cease all activities. A balanced view is needed, such as that given by Susan Bratton:

> Displacement of individual organisms will occur, and is the normal consequence of agriculture. If entire species and ecosystems are lost, however, and diversity is collapsing, we have stolen creation's blessing.[29]

## Secular criticism

Secular critics, together with some Christians, have blasted Christianity for causing most of the world's environmental destruction. Christianity's position toward the environment is labeled as "anthropocentric" because of a misunderstanding of the biblical terms "dominate" and "subdue" (see the chapter by Berry). According to this view, domination led to the careless subjugation of nature, creating our current environmental crisis. Criticisms by secularists are not totally without merit, but by and large they miss the mark. A careful reading of Genesis 1:26 and 28 shows that God gave the command of "dominion" before the Fall. That is, at the time when the Creator saw that his[30] creation was very good (Gen. 1:31).

Dominion and subjugation in this context cannot mean indiscriminate enslavement or reckless abuse of the rest of creation. This is seen in Genesis 2:6–8 and 15, where humankind is created and placed in God's garden to till and keep it. In these and other verses, domination clearly conveys order (till) and care (garden keeper). "Subdue" is the work given to humankind by God to bring order out of chaos. To "inhabit the land that God has given is a gift, transforming it into a house where God can be worshiped"[31] and into an inheritance for their children's children. An image emerges that is near to that of steward or caretaker.

Far from being an anti-environmental religion, Christianity as seen throughout the Bible is very pro-life in all its aspects. Clifford relates that the Bible is unquestionably not anti-ecological. The criticism of Genesis creation stories, voiced by many environmentalists and ecologists, is based on a simplistic literal reading of them and clearly does not represent the core meaning of these texts.[32]

Northcott adds that "the Christian response is at least as significant as that which characterizes the responses of humanists, and of modern philosophers in particular."[33] (See the chapters written by Clifford and Northcott.)

The primary mistake that secular critics make is equating Western civilization with Christianity. The two are interrelated, of course, but also distinct,[34] as demonstrated in the chapter by Harris. The intellectual separation of "nature" from humankind was made possible in the Enlightenment, leading to a false division of knowledge into academic disciplines. Generally, Christianity in the West uncritically absorbed the neo-platonic body-mind, human-nature dualistic mindset of the Enlightenment. Newly emerging technologies succumbed to "mammon," and were not developed as steward's tools to better care for the garden (earth) but as a means of controlling and enslaving the environment. The church should be rightly criticized for abusing its prophetic call on behalf of justice for creation.

Calling to account earth-abusive technologies doesn't mean Christians should be Luddites or intellectual throwbacks to the Palaeolithic age. Scientific techniques and technologies are used by ecologists, geologists, agronomists and other scientists, many of whom are Christian environmentalists. Knowledge and understanding are gifts from God and should be used to glorify the Creator and benefit creation. Significant restoration and renewal of the environment are possible, increasingly so, thanks to scientific and engineering knowledge and methodologies. Perhaps scientists and engineers should be required to make a vow similar to that of medical doctors and promise to do no harm to the patient—in this case, creation.

## Neo-liberals

A criticism from neo-liberal economists, both Christian and secular, castigates environmentalists as a tribe of naysayers. Neo-liberals ask: "Aren't new sources of fuels always found?"; "Isn't per capita food intake up (except in Africa)?"; "Weren't cheaper substitutes found for scarce metals?" In other words, if the price is right, incentives and competition will always provide a cure. Such economic thinking can be a balm to one's conscience. And there's no denying that life is easier for those who have enough dollars to make choices.

Environmentalists respond by claiming that neo-liberal economists are short-sighted. In econometric and other quantitative models,[35] most environmental and social externalities were not included in evaluating the impact of a project through cost/benefit or other techniques in the past.[36] In many countries, these deficiencies are now partially corrected by requiring that shadow prices be placed on non-market items in environmental and social impact statements. In other words, environmentalists are sometimes in the position to use economic tools in acts of stewardship. In other cases this is not so, especially when states ignore their own environmental regulations and laws. Items may still be excluded from evaluation because they remain largely non-quantifiable. Beauty and biodiversity are examples.

### Deep ecologists, Gaians and eco-feminists

Wrongly-perceived Christian positions on the environment are attacked by other critics. Deep ecologists are an eclectic group of humanists, pantheists and some Christians. The usual anthropocentric critique is hurled as an opener. Deep ecologists say that Christians believe humankind reigns supreme above—and distinct from—nature or the environment. They generally envisage all nature (including *Homo sapiens*) as interconnected, and believe that the environment has intrinsic worth apart from that given to it by humankind.

What may shock secular deep ecologists is that most Christian environmentalists agree with these positions. In the words of Bratton, "God has a joy in his creation and purpose in his handiwork that have nothing to do with humankind."[37] On the other hand, Christian deep ecologists may be rightly criticized for the scant role they give to the redeeming role of Christ and the renewing work of the Holy Spirit. God as Creator receives all the attention. Such theologies can border on deism with the result that most self-proclaimed Christian deep ecologists are not Trinitarian in their beliefs.

Going "deeper" still, the concept, notion or religion (depending on the individual) of *Gaia* is encountered. Here, the earth is believed to be a single, self-regulating organism.[38] Everything is interrelated, both ecologically and mystically. All species evolve, and most have become extinct. That's normal. Life continues, though, and will continue with or without humankind. Some Gaia advocates believe that the passing of the human species is very probable, but that's not significant. Other species will flourish and take the place of humankind. The earth (*gaia*) will live. Any environmental destruction, however gross, is an opportunity for other life forms to flourish.

These and similar views place most Gaians beyond Christendom. There may be some who hold to a creator god. But most believers in Gaia see no god that is distinct or separate from nature. Mother Earth or an earth goddess, however expressed, is worshiped.[39] Gaian belief systems are pantheistic or panentheistic New Age religions. The former is the belief that the world as a whole is divine; the latter is a conviction that the earth is part, but not all, of a creator god's being.[40]

Often the interconnectedness theme of deep ecology and Gaia appeals to eco-feminists. Relationships with God, each other and the rest of nature are emphasized, along with belief that cooperation among species is the norm. Competition exists, but within the larger frame of cooperation. "Survival of the fittest" generates the need for cooperation that creates ecological communities, whether human, animal, plant or near-infinite combinations of both. There is a caveat, however. Generally, eco-feminists do not place *Homo sapiens* at the top of the hierarchy of created life. In fact, there isn't a hierarchy, and humankind is certainly not created in the image of God.

The relational, cooperative emphasis is also used in eco-feminist discussions against paternalism. Competitive, low-relational, male-dominated societies are blamed for human-induced ecological disasters. Certainly, within the context of the environmental damage caused by conflict and war, the eco-feminist critique against male authority is convincing.

Interrelationships, cooperation and community in all God's creation resonate with Trinitarian environmentalists, too. What are largely missing from eco-feminist writings that grew out of Christian teachings are the same elements absent in deep ecology linked to Christian traditions: the role of Christ and the work of the Holy Spirit. "Incarnation" is espoused, but the term more often refers to the "body of nature" than to that of Jesus' corporal form.[41] This is surprising, given Jesus' mature and non-paternalistic relationship with women and his challenges to the male-dominated economic and political structures of his day. Redemptive acts of cooperation and community readily flow from Christ's words and deeds.

## Environment care affirmed in the Bible

The Bible is laced with God's love for creation and, by implication, the responsibility for human care of the cosmos. Probably nowhere is the Creator's intimacy with nature expressed more than in the Psalms. The Psalms are a veritable festival of love between God and his created. Psalm 24:1–2 (RSV) reminds us that "the earth is the LORD's and the fulness thereof, the world and all who dwell therein; for he has founded it upon the seas, and established it upon the rivers."

In Psalm 96:11–13 (RSV) we hear a glad creation erupting in praise:

> Let the heavens be glad, and let the earth rejoice;
>> let the sea roar, and all that fills it;
>> let the field exult, and everything in it!
> Then shall all the trees of the wood sing for joy
>> before the LORD, for he comes,
>> for he comes to judge the earth.

That command for all creation to worship the Lord with praise is found again in Psalm 148:7–14 (NAB):

> Praise the LORD from the earth,
>> you sea monsters and all deep waters;
> You lightning and hail, snow and clouds,
>> storm winds that fulfill his command;
>
> You mountains and all hills,
>> fruit trees and all cedars;
> You animals wild and tame,
>> you creatures that crawl and fly;
>
> You kings of the earth and all peoples,
>> princes and all who govern on earth;
> Young men and women too,
>> old and young alike.
>
> Let them all praise the LORD's name,
>> for his name alone is exalted,
>> majestic above earth and heaven.
> The LORD has lifted high the horn of his people;
>> to the glory of all the faithful, of Israel,
>> the people near to their God.

From scriptural passages such as these, it is difficult to comprehend how Christians can have anything but a theocentric worldview of creation and the environment. Perhaps the glory and majesty of the Creator is no better articulated than in Psalm 104:5–13,19–30 (NAB), which states:

> You fixed the earth on its foundation,
>  never to be moved.
> The ocean covered it like a garment;
>  above the mountains stood the waters.
> At your roar they took flight;
>  at the sound of your thunder they fled.
> They rushed up the mountains,
>  down the valleys to the place you had fixed for them.
> You set a limit they cannot pass;
>  never again will they cover the earth.
>
> You made springs flow into channels
>  that wind among the mountains.
> They give drink to every beast of the field;
>  here wild asses quench their thirst.
> Beside them the birds of heaven nest;
>  among the branches they sing.
> You water the mountains from your palace;
>  by your labor the earth abounds...
>
> You made the moon to mark the seasons,
>  the sun that knows the hour of its setting.
> You bring darkness and night falls,
>  then all the beasts of the forest roam abroad.
> Young lions roar for prey;
>  they seek their food from God.
> When the sun rises, they steal away
>  and rest in their dens.
> People go forth to their work,
>  to their labor till evening falls.
>
> How varied are your works, LORD!
>  In wisdom you have wrought them all;
>  the earth is full of your creatures.
> Look at the sea, great and wide!
>  It teems with countless beings,
>  living things both large and small.
> Here ships ply their course;
>  here Leviathan, your creature, plays.

All of these look to you
  to give them food in due time.
When you give to them, they gather;
  when you open your hand, they are well filled.
When you hide your face, they are lost.
  When you take away their breath, they perish
  and return to the dust from which they came.
When you send forth your breath, they are created,
  and you renew the face of the earth.

In Isaiah there are passages rich with imagery that stress God's majesty, and by comparison, humankind's:

Who has cupped in his hand the waters of the sea,
  and marked off the heavens with a span?
Who has held in a measure the dust of the earth,
  weighed the mountains in scales
  and the hills in a balance?
Who has directed the spirit of the LORD,
  or has instructed him as his counselor?
Whom did he consult to gain knowledge?
  Who taught him the path of judgment,
  or showed him the way of understanding?
Behold, the nations count as a drop in the bucket,
  as dust on the scales;
  the coastlands weigh no more than powder.
Lebanon would not suffice for fuel,
  nor its animals be enough for holocausts.
Before him all the nations are as nought,
  as nothing and void he accounts them…

Do you not know? Have you not heard?
  Was it not foretold you from the beginning?
  Have you not understood? Since the earth was founded
He sits enthroned above the vault of the earth,
  and its inhabitants are like grasshoppers;
He stretches out the heavens like a veil,
  spreads them out like a tent to dwell in.
He brings princes to nought
  and makes the rulers of the earth as nothing.
Scarcely are they planted or sown,
  scarcely is their stem rooted in the earth,

When he breathes upon them and they wither,
  and the stormwind carries them away like straw.

To whom can you liken me as an equal?
  says the Holy One.
Lift up your eyes on high
  and see who has created these:
He leads out their army and numbers them,
  calling them all by name.
By his great might and the strength of his power
  not one of them is missing! (Isa. 40:12–17, 21–26, NAB)

Isaiah also brings us one of the greatest prophecies of hope through the redemption of the world in Jesus Christ and the renewing of creation by the Holy Spirit:

Lo, I am about to create new heavens
  and a new earth;
The things of the past shall not be remembered
  or come to mind.
Instead, there shall always be rejoicing and happiness
  in what I create;
For I create Jerusalem to be a joy
  and its people to be a delight;
I will rejoice in Jerusalem
  and exult in my people.
No longer shall the sound of weeping be heard there,
  or the sound of crying;
No longer shall there be in it
  an infant who lives but a few days,
  or an old man who does not round out his full lifetime;
He dies a mere youth who reaches but a hundred years,
  and he who fails of a hundred shall be thought accursed.
They shall live in the houses they build,
  and eat the fruit of the vineyards they plant;
They shall not build houses for others to live in,
  or plant for others to eat.
As the years of a tree, so the years of my people;
  and my chosen ones shall long enjoy
  the produce of their hands.
They shall not toil in vain,
  nor beget children for sudden destruction;

> For a race blessed by the LORD are they
>   and their offspring.
> Before they call, I will answer;
>   while they are yet speaking, I will hearken to them.
> The wolf and the lamb shall graze alike,
>   and the lion shall eat hay like the ox
>   (but the serpent's food shall be dust).
> None shall hurt or destroy
>   on all my holy mountain, says the LORD.
>     (Isa. 65:17–25, NAB)

Scripture in the New Testament confirms Christ's lordship over creation (Luke 8:23–24, John 1:1–4). Yet New Testament passages directly related to creation, environment or stewardship are few. We must remember that the New Testament is both a continuation and fulfillment of the Old. The overall focus of the New Testament is on the kingdom of God through belief in the lordship of Christ. Trinitarian theology announces Jesus as co-Creator (John 1:1–3, Col. 1:15–16) and Redeemer of humans and all creation. In the words of Paul:

> For in him all the fulness of God was pleased to dwell, and through him to reconcile to himself all things, whether on earth or in heaven, making peace by the blood of his cross. (Col 1:19–20, RSV)

God's re-creation power is revealed through the Holy Spirit (Matt 23:19, Mark 13:4, Luke 3:22, 2 Cor. 13:13). In the words of Christ as found in John 14:26 (RSV):

> But the Counselor, the Holy Spirit, whom the Father will send in my name, he will teach you all things, and bring to your remembrance all that I have said to you.

## Closing

A biblically-based environmental theology is founded on three principles. First, The Bible is a history of God's work, not human achievement. Clearly, Christian environmental theology is theo-centric, not anthropocentric. God is seen in the role of creator, redeemer, and restorer of all creation. We worship a triune God. The words and work of Christian environmentalists should reflect this belief.

Second, humankind and the rest of the cosmos are part of the created order. Yet humans are unique in that they are made in the image of God. The primary task of humankind is to glorify God and take joy in his presence.

Third, humans were given the responsibility to care for God's garden, the earth. In this respect, "dominion" and "subjugation" are best seen through the lens of stewardship. But a steward must not be seen as a co-creationist with God. We are to respect the rest of nature, to bless and be blest in so doing.

---

[1]  Hans Schwarz, "A Critique of Christendom's Relation to Land with Direction towards a Christian Land Ethic," Forum Papers, Au Sable Institute, Madison, Wis., 1987, 18.

[2]  "End Time" Christians tend to emphasise apocalyptic passages in Ezekiel, Daniel, and Revelation. Christian environmentalists lean more on Genesis, Deuteronomy, Psalms, Isaiah, the Gospels, Romans, and Colossians.

[3]  Professor of Theology and Culture at Eastern Baptist Theological Seminary and a founder of Evangelicals for Social Action.

[4]  Ronald J. Sider, "Biblical Foundations for Creation Care," in *The Care of Creation*, ed. R.J. Berry, (Leicester, UK: Inter-Varsity Press, 2000), 43.

[5]  Israelites worshiped Baal simultaneously with the Lord. Yahweh was the high and feared god, while Baal served as the lesser but seemingly more practical and down to earth god of fertility. See Paul G. Hiebert, R. Daniel Shaw, and Tite Tienou, *Understanding Folk Religion: A Christian Response to Popular Beliefs and Practices* (Grand Rapids, MI: Baker Books, 1999), 48.

[6]  Calvin B. DeWitt, "Creation's Environmental Challenge to Evangelical Christianity," in *The Care of Creation*, ed. R.J. Berry, (Leicester, UK: Inter-Varsity Press, 2000), 71–72.

[7]  Susan Power Bratton, "A Fierce Green Fire Dying: Christian Land Ethics and Wild Nature," Forum Papers, Au Sable Institute, Madison, Wis., 1987, 11–12.

[8]  Michael S. Northcott, *The Environment and Christian Ethics* (Cambridge, UK: Cambridge University Press, 1996), 208.

[9]  Joseph Marie Plunkett, "I See His Blood upon the Rose," available at http://poetry.elcore.net/CatholicPoets/Plunkett/EaL_Title.html, early twentieth century.

[10] Quoted in Northcott, loc. cit., 228.

[11] Anne M. Clifford, "Foundations for a Catholic Ecological Theology of God," in *And God Saw that It Was Good: Catholic Theology and the Environment*, eds. Drew Christiansen and Walter Grazer, (Washington, D.C.: United States Catholic Conference, 1996), 27.

[12] United States Catholic Conference, found in Christiansen and Grazer, loc. cit., 235.

[13] Christensen and Grazer, "Introduction," loc. cit., 14.

[14] Agni Vlavianos-Arvantis, "Biopolitics—The Bioenvironment: Biocentric Values and Ethics for the Next Millennium," Ecumenical Patriarch of Constantinople, *Orthodox World News* Web site, 1994, 2.

15 Robert Flanagan, "Orthodoxy and the Environment," *Jacob's Well* Web site, The Orthodox Church in America, Diocese of New York and New Jersey, Winter 1996, 2.

16 Fyodor Dostoyevsky, "Love Reveals the Mysteries of Creation," quoted in Susan Bratton, "Christian Care for Creation," *Green Cross*, Vol. 1, No. 2, Winter 1995, 11.

17 Ibid.

18 Anestis G. Keselopoulos, *Man and the Environment: A Study of St Symeon the New Theologian*, translated by Elizabeth Theokritoff, (Crestwood, NY: St Vladimir's Seminary Press, 2001), 110.

19 Metropolitan John of Pergamon, "Orthodoxy and Ecological Problems: A Theological Approach," Ecumenical Patriarch of Constantinople, *Orthodox World News* Web site, 1994, 5.

20 Separating denominations between conciliar and evangelical is a crude division. Certainly many mainline parishioners are evangelical in their core beliefs. Likewise, an increasing number of evangelical Christians are active in their promotion of social justice.

21 Initial support from the Tear Fund and World Vision.

22 An example is "Eco-Myths," by David N. Livingstone, Calvin B. DeWitt, and Loren Wilkinson, *Christianity Today*, 25 June 2001 (Web site).

23 Ronald J. Sider, "Biblical Foundations for Creation Care," in Berry, loc. cit., 43.

24 Michael Northcott, "The Spirit of Environmentalism," in Berry, loc. cit., 168.

25 Calvin B. DeWitt, *The Just Stewardship of Land and Creation* (Grand Rapids, MI: Reformed Ecumenical Council, 1996), 42.

26 See Ronald J. Sider, "Tending the Garden without Worshiping It," in *The Best Preaching on Earth: Sermons on Caring for Creation*, ed. Stan L. LeQuire, (Valley Forge, PA: Judson Press and Evangelical Environmental Network, 1996), 31.

27 Reduction of bio-diversity is by no means the only violent act on the environment perpetuated by "bad stewards." The litany of evils include, among others, global warming, reduction of ozone layer, soil erosion and salization, deforestation, acid rain, water pollution by industries, farm run-off and human excrement, and over fishing.

28 Quoted in Web of Creation, "Section 5. Public Ministry and Advocacy, Biodiversity," in Transforming Faith-based Communities for a Sustainable World (Web of Creation Web site, c. 1999), 7.

29 Bratton, loc. cit., 32.

30 Keeping with convention, masculine pronouns for God will be used in this report. While I believe that the Creator has no need of gender, or possesses both genders, depending on one's perspective, the term "Father God" assures that we stay out of the trap of pantheism where the image of "Mother" may lead us.

31 Clifford, loc. cit., 27.

32 Ibid., 41.

33 Northcott, *Environment and Christian Ethics*, loc. cit., 124.

34 By analogy, we could say that Buddhism encourages crass materialism and the blatant destruction of ecosystems from an examination of Japanese urban areas or the operations of Japanese timber companies in Asia and Latin America.

35 One definition of economics is the science of rationing scarce resources. "Scarce" means that a value (quantity) may be placed on something, thereby allowing commodities or activities to be quantitatively analyzed.

36 Externalities may be good (valuable) or bad (noxious) activities that are not accounted for by parties causing or benefiting by the externality. An example is a highway improvement where commuters and merchants benefit at the expense of noise and air pollution to nearby residences.

37 Bratton, loc. cit., 22.

38 According to Wilkinson, Gaians believe that the unstable chemistry of the atmosphere provides a stable environment because the earth (Gaia) adjusts, adopts, and otherwise compensates for changes. Wilkinson, Loren, "Gaia Spirituality: A Christian Perspective," in "Evangelicals and the Environment: Theological Foundations for Christian Environmental Stewardship," ed. J. Mark Thomas, *Evangelical Review of Theology*, Vol. 17, No. 2, April 1993, 178.

39 It's fair to say that Gaia as a belief system is held by few scientists. It's also fair to say that the cosmic constructs of some "process" theologians are essentially Gaian.

40 Definitions after Northcott, loc. cit., 86.

41 Robert B. Fowler, *The Greening of Protestant Thought* (Chapel Hill: University of North Carolina Press, 1995), 131.

# An Evangelical Declaration on the Care of Creation[1]

"The earth is the Lord's, and the fullness thereof" (Psalm 24:1).

*As followers of Jesus Christ, committed to the full authority of the Scriptures, and aware of the ways we have degraded creation, we believe that biblical faith is essential to the solution of our ecological problems.*

- Because we worship and honour the Creator, we seek to cherish and care for the creation.

- Because we have sinned, we have failed in our stewardship of creation. Therefore we repent of the way we have polluted, distorted, or destroyed so much of the Creator's work.

- Because in Christ God has healed our alienation from God and extended to us the first fruits of the reconciliation of all things, we commit ourselves to working in the power of the Holy Spirit to share the Good News of Christ in word and deed, to work for the reconciliation of all people in Christ, and to extend Christ's healing to suffering creation.

- Because we await the time when even the groaning creation will be restored to wholeness, we commit ourselves to work vigorously to protect and heal that creation for the honor and glory of the Creator—whom we know dimly through creation, but meet fully through Scripture and in Christ.

*We and our children face a growing crisis in the health of creation in which we are embedded, and through which, by God's grace, we are sustained. Yet we continue to degrade that creation.*

- These degradations of creation can be summed up as: (1) land degradation; (2) deforestation; (3) species extinction;

(4) water degradation; (5) global toxification; (6) the alter-
ation of atmosphere; (7) human and cultural degradation.

- ◦ Many of these degradations are signs that we are pressing
  against the finite limits God has set for creation. With continued population growth, these degradations will become
  more severe. Our responsibility is not only to bear and nurture children, but to nurture their home on earth. We respect the institution of marriage as the way God has given
  to ensure thoughtful procreation of children and their nurture to the glory of God.

- ◦ We recognise that human poverty is both a cause and a consequence of environmental degradation.

*Many concerned people, convinced that environmental problems
are more spiritual than technological, are exploring the world's
ideologies and religions in search of non-Christian spiritual
resources for the healing of the earth. As followers of Jesus Christ,
we believe that the Bible calls us to respond in four ways:*

- ◦ First, God calls us to confess and repent of attitudes which
  devalue creation, and which twist or ignore biblical revelation to support our misuse of it. Forgetting that "the earth is
  the Lord's," we have often simply used creation and forgot-
  ten our responsibilty to care for it.

- ◦ Second, our actions and attitudes towards the earth need
  to proceed from the centre of our faith, and be rooted in the
  fullness of God's revelation in Christ and the Scriptures. We
  resist both ideologies which would presume the Gospel has
  nothing to do with the care of non-human creation and also
  ideologies which would reduce the Gospel to nothing more
  than the care of that creation.

- ◦ Third, we seek carefully to learn all that the Bible tells us
  about the Creator, creation, and the human task. In our life
  and words we declare that full good news for all creation
  which is still waiting "with eager longing for the revealing of
  the children of God" (Rom. 8:19).

- ◦ Fourth, we seek to understand what creation reveals about
  God's divinity, sustaining presence, and everlasting power,
  and what creation teaches us of its God-given order and the
  principles by which it works.

*Thus we call on all those who are committed to the truth of the Gospel of Jesus Christ to affirm the following principles of biblical faith, and to seek ways of living out these principles in our personal lives, our churches, and society:*

- ⚘ The cosmos, in all its beauty, wildness, and life-giving bounty, is the work of our personal and loving Creator.

- ⚘ Our creating God is prior to and other than creation, yet intimately involved with it, upholding each thing in its freedom, and all things in relationships of intricate complexity. God is transcendent, while lovingly sustaining each creature; and immanent, while wholly other than creation and not to be confused with it.

- ⚘ God the Creator is relational in very nature, revealed as three persons in One. Likewise, the creation which God intended is a symphony of individual creatures in harmonious relationship.

- ⚘ The Creator's concern is for all creatures. God declares all creation "good" (Gen. 1:31); promises care in a covenant with all creatures (Gen. 9:9–17); delights in creatures which have no human apparent usefulness (Job 39–41); and wills, in Christ, "to reconcile all things to himself" (Col. 1:20).

- ⚘ Men, women and children, have a unique responsibility to the Creator; at the same time we are *creatures*, shaped by the same processes and embedded in the same systems of physical, chemical, and biological interconnections which sustain other creatures.

- ⚘ Men, women and children, created in God's image, also have a unique responsibilty for creation. Our actions should both sustain creation's fruitfulness and preserve creation's powerful testimony to its Creator.

- ⚘ Our God-given, stewardly talents have often been warped from their intended purpose: that we know, name, keep and delight in God's creatures; that we nourish civilization in love, creativity and obedience to God; and that we offer creation and civilisation back in praise to the Creator. We have ignored our creaturely limits and have used the earth with greed, rather than care.

↬ The earthly result of human sin has been a perverted stewardship, a patchwork of garden and wasteland in which the waste is increasing. "There is no faithfulness, no love, no acknowledgement of God in the land . . . Because of this the land mourns, and all who live in it waste away"(Hosea 4:1,3). Thus, one consequence of our misuse of the earth is an unjust denial of God's created bounty to other human beings, both now and in the future.

↬ God's purpose in Christ is to heal and bring to wholeness not only persons but the entire created order. "For God was pleased to have all his fullness dwell in him, and through him to reconcile to himself all things, whether things on earth or things in heaven, by making peace through his blood shed on the cross" (Col. 1:19–20).

↬ In Jesus Christ, believers are forgiven, transformed and brought into God's kingdom. "If anyone is in Christ, there is a new creation" (2 Cor. 5:17). The presence of the kingdom of God is marked not only by renewed fellowship with God, but also by renewed harmony and justice between people, and by renewed harmony and justice between people and the rest of the created world. "You will go out with joy and be led forth in peace; the mountains and the hills will burst into song before you, and all the trees of the field will clap their hands" (Isa. 55:12).

*We believe that in Christ there is hope, not only for men, women and children, but also for the rest of creation which is suffering from the consequences of human sin.*

↬ Therefore we call upon all Christians to reaffirm that all creation is God's; that God created it good; and that God is renewing it in Christ.

↬ We encourage deeper reflection on the substantial biblical and theological teaching which speaks of God's work of redemption in terms of the renewal and completion of God's purpose in creation.

↬ We seek a deeper reflection on the wonders of God's creation and the principles by which creation works. We also urge a careful consideration of how our corporate and individual actions respect and comply with God's ordinances for creation.

↬ We encourage Christians to incorporate the extravagant creativity of God into their lives by increasing the nurturing role of beauty and the arts in their personal, ecclesiastical and social patterns.

↬ We urge individual Christians and churches to be centres of creation's care and renewal, both delighting in creation as God's gift, and enjoying it as God's provision, in ways which sustain and heal the damaged fabric of the creation which God has entrusted to us.

↬ We recall Jesus' words that our lives do not consist in the abundance of our possessions, and therefore we urge followers of Jesus to resist the allure of wastefulness and over-consumption by making personal lifestyle choices that express humility, forbearance, self restraint and frugality.

↬ We call on Christians to work for godly, just, and sustainable economies which reflect God's sovereign economy and enable men, women and children to flourish along with all the diversity of creation. We recognise that poverty forces people to degrade creation in order to survive; therefore we support the development of just, free economies which empower the poor and create abundance without diminishing creation's bounty.

↬ We commit ourselves to work for responsible public policies which embody the principles of biblical stewardship of creation.

↬ We invite Christians—individuals, congregations and organisations—to join with us in this evangelical declaration on the environment, becoming a covenant people in an ever-widening circle of biblical care for creation.

↬ We call upon Christians to listen to and work with all those who are concerned about the healing of creation, with an eagerness both to learn from them and also to share with them our conviction that the God whom all people sense in creation (Acts 17:27) is known fully only in the Word made flesh in Christ the living God, who made and sustains all things.

↝ We make this declaration knowing that until Christ returns
to reconcile all things, we are called to be faithful stewards
of God's good garden, our earthly home.

---

1  The *Declaration* was launched in 1994 as one of the first products of the Evangelical
Environment Network (EEN), and has since been endorsed by several hundred church
leaders throughout the world. For more information regarding the Declaration and
the origin of the EEN, please see *The Care of Creation: Focusing Concern and Action*,
ed. R.J. Berry, (Inter-Varsity Press: England, 2000).

World Vision

# Other Titles from World Vision Publications

### COMPLEX HUMANITARIAN EMERGENCIES
### Lessons from Practitioners
*Mark Janz*
*& Joann Slead, editors*
In this volume, experienced practitioners address the question of how we can respond appropriately to CHEs by linking conceptual and theoretical thinking to practical application at the grassroots level.

288 pp.
2000
Y-007
$24.95

### PROTECTING CHILDREN
### A Biblical Perspective on Child Rights
From its rich history of service to those who are exploited, abused and traumatized, World Vision explains the rights of children from a biblical perspective and how that integrates with standards set by the UN Convention on the Rights of the Child.

38 pp
2002
Free

### WORKING WITH THE POOR
### New Insights and Learnings from Development Practitioners
*Bryant L. Myers, editor*
Here, development practitioners from around the world struggle to overcome the Western assumption that the physical and spiritual realms are separate and distinct from one another in answering the question, How do Christian practitioners express authentically holistic transformational development?

192 pp.
1999
Y-002
$16.95

### MASTERS OF THEIR OWN DEVELOPMENT
### PRSPs and the Prospects for the Poor
*Alan Whaites, editor*
"This report by my colleagues from World Vision has outlined many of the problems that have been encountered with PRSPs [Poverty Reduction Strategy Papers] since 1999. Their analysis is all the more interesting because World Vision has tried to support the idea of PRSPs from the start. It is clear from their research that even those who have stood by PRSPs are finding the experience increasingly frustrating."
—Ann Pettifor, from the introduction

288 pp.
2002
Y-022
$29.95

### WALKING WITH THE POOR
### Principles and Practices of Transformational Development
*Bryant L. Myers*
Drawing on theological and biblical resources, secular development theory and work done by Christians among the poor, Myers develops a theoretical framework for transformational development and provides cutting-edge tools for those working alongside the poor.

288 pp.
1999
Y-008
$21.95

## To order call 1-800-777-7752 (US) • 1-626-301-7720
•wvpp@wvi.org •